T0199265

making mathematics with needlework

making mathematics with needlework

ten papers and ten projects

edited by

SARAH-MARIE BELCASTRO

CAROLYN YACKEL

CRC Press
Taylor & Francis Group
Boca Raton London New York

CRC Press is an imprint of the
Taylor & Francis Group, an **informa** business

AN A K PETERS BOOK

Editorial, Sales, and Customer Service Office

CRC Press
Taylor & Francis Group
6000 Broken Sound Parkway NW, Suite 300
Boca Raton, FL 33487-2742

First issued in hardback 2019

Library of Congress Cataloging-in-Publication Data

Making mathematics with needlework : ten papers and ten projects / edited by Sarah-Marie Belcastro, Carolyn Yackel.
p. cm.
Includes bibliographical references and index.
ISBN 978-1-56881-331-8 (alk. paper)
1. Needlework. 2. Fiberwork. 3. Mathematics. I. Belcastro, Sarah-Marie. II. Yackel, Carolyn.

TT751.M27 2007
510.71–dc22

2007022382

Visit the Taylor & Francis Web site at
http://www.taylorandfrancis.com

and the CRC Press Web site at
http://www.crcpress.com

To my grandmothers, who were excellent needleworkers
but didn't know they were mathematicians.
—Carolyn Yackel

To my mother, who knits, and my father, who sews—both needleworkers
who ought to know by now that they're mathematicians.
—sarah-marie belcastro

CONTENTS

ACKNOWLEDGMENTS

As with any book, there are a large number of people who contributed to this volume's completion. Obviously we are grateful to the chapter authors, for without them there would literally have been no book! Also in this category are Sharon Frechette of the College of the Holy Cross for proposing the chapter on cabling and Daina Taimina of Cornell University for writing articles and producing objects that inspired us to create the chapter on hyperbolic pairs of pants. (Keep an eye out for her forthcoming book on *Exploring Non-Euclidean Geometry Through Crochet*.) We thank the American Mathematical Society for approving our request to hold the Special Session in Mathematics and Mathematics Education in Fiber Arts, which took place at the January 2005 Joint Mathematics Meetings in Atlanta, Georgia. We also thank the session speakers, many of whom are chapter authors. We are especially grateful to Alice Peters for agreeing to publish the book and for helping us to figure out how to do it, to Mandy Love for keeping us on track in Fall 2006, Tom Hull for assisting sarah-marie with diagrams and talking through some of the mathematics, Sean Kinlin and Tom Hull for listening to sarah-marie despair over the book, Erna "Mom" Yackel for serving as a limitless resource and sounding board for Carolyn with respect to mathematics education, and Creighton Rosental for feeding Carolyn and keeping her sane.

Thanks to our project testers, Joan "Mom" Belcastro, Ann Black, Jane Carder, Charlotte Davis, Emily Gavin, Ashley Hatfield, Zia Marek-Loftus (who tested four projects), Kim Plofker, Jeri Riggs, Em Schoch, Sue Sierra, and Aunt Ida Thornton. Thanks also to Smith mathematics students Constance Baltera, Natasha Resendes, Liz Sullivan, and Angela Turner, who participated in the Fortunatus's Hat photo shoot but aren't pictured in the book (and Portia Parker and Madison Stuart, who are) and to additional models Arzachel, Frank "Daddy" and Joan "Mom" Belcastro, Rachel Brown, Ashley Hatfield, Zia Marek-Loftus, Michela Rowland, Hope Stansfield, and Patricia Williams. sarah-marie is particularly grateful to those who provided us with measurements for the hyperbolic pants, including Julie Brand Asbornsen (who measured 13 children), Tim and Elaine Comar, Dana Rowland, Erika King via Ruth Haas, Cheri Kraske (who measured a one-day-old grandchild), Alice Peters, Kate Queeney, Elizabeth Wilmer, and Dan Zook.

The community of fiber artists with a significant interest in mathematics and mathematicians who think about fiber arts would be poorer without people like

* Lara Neel of the *Math4Knitters* blog and podcast;

* Brenda Dayne, who wrote the "Geek Chic" article in *Interweave Knits* that seemed to take the world by storm (and who now does the popular podcast *Cast On*);

* Debbie New, who was a microbiologist before she became famous for her (frequently mathematical) knitting;

* Lesley Perg of "thomasina's Guide to Geeky Knitting," who maintains a web index, broken out by discipline, of scientifically-related knitting;

* Ruth Stewart of the *Impulse of Delight* website who is a physician, a knitter, and a jeweler who knits and crochets with wire (and creates Möbius bands, trefoil knots, hyperbolic orchids...);

* Mark Shoulson, Claire Irving, and Nate Berglund, who knit and crochet a variety of embeddings of topological objects;

* Hinke Osinga, who created the crocheted Lorenz manifold;

* Miss Lime and Miss Violet (of *Lime & Violet* podcast fame) for liking algebra and spacial relations, respectively, and for saying loudly both that girls should not be discouraged from mathematics and that mathematics is necessary for knitting;

★ the hundred or so members of our mathematical fiber arts email list, most of whom have attended Knitting Networks at national mathematics conferences. (There's not enough room to list all of your names here.)

We are grateful to all of these folks for contributing to the confluence of mathematics and fiber arts.

From Chapter Authors. Joshua Holden wants to thank Lana for mathematical and fiberological support and Richard and Andrea for generally making life better. Also, Susan Wildstrom helped a great deal with his "Teaching Ideas" section. Amy F. Szczepański wants to thank Jim Conant for help with revisions. D. Jacob Wildstrom wants to thank Mary Pat Campbell for inspiring his Sierpinski crochet work and graciously allowing the reproduction of her design methods in his chapter. Thanks also go to the brothers of Epsilon Theta Fraternity, who taught him to crochet, and to his mother, who valiantly attempted to teach him to knit. Susan Goldstine wants to thank her Amherst College First-Year Seminar Program students for inspiring her to make her first Fortunatus's Purse and her colleagues at St. Mary's College of Maryland for supporting her expository activities, especially this work.

INTRODUCTION

an overview of
mathematics and fiber arts

SARAH-MARIE BELCASTRO
CAROLYN YACKEL

1   Welcome!

Welcome to the first book to contain both mathematics papers and fiber arts project instructions. This volume brings together eight mathematicians to present a wide variety of mathematics research and work in mathematics education related to the fiber arts. It is structured to be of interest to mathematicians, mathematics educators, and crafters; every chapter has an overview as well as sections on mathematics and mathematics education and the chapter's project. All readers will be able to understand the overview sections, as they include introductions to the various fiber arts as well as lay summaries of the mathematical content. While the mathematics sections are written for mathematicians, the authors have made a special effort to make their work accessible to lay readers by providing definitions of mathematical terms and many diagrams. Instructors at all levels—ranging from elementary school through graduate programs—will find ways to use fiber arts in their classrooms in the teaching ideas sections. Not every topic is suitable for every class, of course, but there really is something for everyone. Finally, the project sections are written for crafters, so that our non-mathematician readers can have a tangible experience with mathematical concepts.

At the same time, this is a very specialized volume. First, there are many fiber arts that are not represented (for example, tatting, felting/fulling, and weaving) but are associated with interesting mathematics. The selection of fiber arts represented here is a direct result of the book's origin—it grew out of the American Mathematical Society Special Session in Mathematics and Mathematics Education in Fiber Arts, held at the January 2005 Joint Mathematics Meetings in Atlanta, Georgia. The needlework discussed in the session (knitting, crochet, cross-stitch, embroidery, and sewing/quilting) determined what we could include in this book. Second, for each fiber art there is a wide range of mathematics

and mathematics education to be discussed, but only a sampling is shown here. Again, this is somewhat an artifact of what work mathematicians chose to present in the Special Session. However, it is also true that there has been very little published work available in mathematics and fiber arts—aside from quilting and weaving, that is. On pages 7–10 we give bibliographies of all published material on mathematics or mathematics education and fiber arts known to the editors as this volume went to press.

The rest of this introduction contains a brief history of mathematical work on fiber arts, a discussion of fiber arts in mathematics education, and an overview of present and future mathematics research related to fiber arts.

2   A Short History

Elementary mathematics in the form of arithmetic and geometry has been used in fiber arts throughout human history. While that's not so interesting to professional mathematicians, in terms of mathematics education it's very interesting. For students who have prior experience with fiber arts, we can make explicit the mathematics they already use as a way to grow their experiences. Further, teaching various fiber arts to children can be a vehicle for introducing or reinforcing many mathematical concepts. Mary Harris created a sequence of mathematics teaching activities based on textile arts. Unfortunately, this work, called *Cabbage*, seems to be unavailable either in print or electronically. A couple of examples from the collection in her book *Common Threads* [14] highlight the significant mathematics present in simple practical issues in fiber arts.

That said, what sort of work *has* been done on mathematics and fiber arts? This miniature history is based only on published work—and only on those that we have been able to find—so it won't be surprising if readers know of sources we have missed. (Please contact us if you have additional information!)

There has been more published on the mathematics of weaving than on mathematics and any other fiber art—and apparently more than on mathematics and *all* other fiber arts. This began with three papers by E. Lucas published between 1867 and 1911 [45–47], followed by a paper of S. A. Shorter in 1911 [52] and a section of a more general work on pattern design by H. J. Woods in 1935 [54]. As far as we know, nothing more was done in the following 45 years, until Grünbaum and Shephard wrote in 1980 about the mathematics of weaving in satins and twills [30]. The approach is similar to their work on tilings and patterns. Their article spawned many other papers on the topic in the 1980s and early 1990s, including four by C. R. J. Clapham [26–29], two by Jean Pedersen [48, 49], eleven by Janet Hoskins [34–44] involving five varying collaborators, two by Richard Roth [50, 51], and three more by Grünbaum and Shephard themselves [31–33]. The *Mathematical Reviews* review (MR0674144) by G. Ewald of an early article by Hoskins [34] mentions that the mathematics of weaving had been of great interest for nearly a decade. Very recently, R. S. D. Thomas (a collaborator of Hoskins) released a preprint on isonemal prefabrics [53].

The history of mathematical publications on other fiber arts is much more recent. It begins in 1971 with a paper on knitting topological surfaces by Miles Reid [20], followed nearly fifteen years later by a quartet of papers in secondary mathematics education journals on crochet, quilting, sewing, and knitting [5, 7, 13, 21] and a blurb on random processes in knitting in a popular science magazine [12]. Another decade-plus passed before the current wave of publications began. In knitting, we have papers by Dan Isaksen and Al Petrofsky [17] and Claire Irving [16] on topological surfaces; in crochet, there are papers by David Henderson and Daina Taimina [15] and Hinke Osinga and Bernd Krauskopf [19] on geometric surfaces; in quilting, Gwen Fisher [8–10] and Mary

Williams [23, 24] have written about depictions of mathematical concepts and objects; and Therese Biedl, John Horton, and Alejandro Lopez-Ortiz did work on thread optimization in cross-stitch [4].

A computer search of the literature yields basically nothing. MathSciNet (the comprehensive database of mathematics research papers) lists one article about knitting from 1999 [11], and it's really about notation for drawing knitted fabric. A few papers deal with the "knitting ansatz" of braid theory (for basic information on braid groups, see Chapter 8), which has nothing to do with fiber arts. MathSciNet also lists many promising references to crochet, except that "crochet" is French for "bracket" and so they're really all about Lie theory. (The two exceptions are [15] and [19].) Even quilting gets short shrift. There are two articles [6, 22] whose titles refer to quilts, but they are about dissections of rectangles into squares; neither appears to have been inspired by actual quilts. Conway and Hsu have named a group-theoretical concept a "quilt." And that's it. Completely. Nothing about tatting, or cross-stitch, or lace-making. So, mathematicians and fiber artists, let's get going! The mathematical world is ripe for our work. The next two sections provide some ideas for how to proceed.

3   Fiber Arts and Mathematics Education

Fiber arts problems and applications can be useful in a variety of classroom teaching styles. They can be used as applications by teachers who prefer to present theory first and give applications afterwards, as described below. They can also be used as motivation for development of mathematics, including development from a constructivist perspective. Indeed, through very careful design one can even create lessons that conform to Gravemeijer's notion of realistic mathematics education. (See [56, pp. 82–83].)

One common response of mathematics educators to students desiring to understand the utility of the mathematics being taught in the classroom is to assign applications problems. However, as fascinating as we may find the fact that it is possible to calculate the escape velocity from the moon, arguably a student may not find that applicable to his or her life. Is he or she really going to launch something at that velocity off of the face of the moon in order to send it into orbit? Likely not. How about the classic linear constraint of butter versus guns? This oversimplification of military spending is unsatisfactory to our students. The fact is that applications problems are often not grounded in the reality of students' lives. Further, they are usually so complicated that only a trivial version can be presented. With problems arising from fiber arts, often both difficulties are avoided. At least some portion of the audience can relate to the scenario, and with careful problem selection, the problems can be real, complete, and at the appropriate level.

Some instructors like to begin with a problem, real-world or otherwise, to motivate the development of a topic in mathematics. Fiber arts problems are wonderful venues for this, because they are real-world problems that real people actually want answered. Through successive passes at the problem, during each of which the scenario is formulated more and more mathematically, teachers model the process of mathematizing the situation, using Freudenthal's terminology [9, pp. 30–31]. (For a detailed explanation of the term, see [56].) At some point the problem is stated mathematically. Through this process the teacher has motivated the topic development. However, if the instructor wishes students to take the lead, careful planning of questions leading them along a learning trajectory [59] may help students to successfully transform the scenario into a relevant mathematical problem on their own.

At some stage, student autonomy in this regard should be a goal, because without the ability to reframe an everyday situation as mathematical, a person will never be mathematically fully functioning. That is, for math to be of maximal use to a person, that person must have a belief in the utility of viewing situations from a mathematical perspective and the skills and initiative to make successive translations of an everyday problem that is not presented in a mathematical format into the underlying essential question, which *is* mathematically stated. Of course, not all everyday scenarios can be developed into math problems, so the adroit user of mathematics must also have a sense of which questions are at heart mathematical and which are not.

This leads us to the question: Is it possible to teach another person to mathematize? This question is similar to, but not the same as, the old question of whether or not one can teach problem solving. Of course, teachers can present techniques and strategies. The most famous endeavor in this regard is Polya's *How to Solve It* [58]. Yet, creativity plays a large role, and often all we can do as instructors is to model that behavior ourselves. Indeed, the brilliance of the authors contributing to this book was to recognize which parts of mathematics applied to their arts and precisely how that mathematics should be applied. Throughout the book, the authors link the details of particular mathematics with a particular fiber art, focussing alternately on an aspect of the craft and an aspect of the mathematics, refining the mathematics to better describe the situation, and so forth. All of these activities and the development of general mathematical explanations and formulas are included in the term "mathematizing."

Nonetheless, the authors noticed the mathematics they did in their particular fiber arts because they had prior knowledge of that mathematics. (For more on *noticing*, see [57].) Therefore, it is through that knowledge that the fiber art became a mode of expression of the mathematics for that mathematician. Alternatively, a fiber artist who has no knowledge of the mathematics in question could make the same piece and admire the final composition, but for this second artist, the piece would not be expressive of the mathematics that it so beautifully exhibits for the mathematician/artist. So, the mathematical value could be said to be in the eye or the mind of the beholder.

In that case, what is the educational value of mathematical fiber arts? We claim several possible positive outcomes.

★ Actually participating in the kinesthetic experience of making a fiber object will create a different type of mental link than can be achieved by reading or listening to a word problem. This will make any problem that arises very real and immediate to the students.

★ Having experiences solving real-world problems posed in a non-mathematical format, by going through the process of restating the problems as mathematical and following through with a mathematical solution, may help students build confidence in their abilities to do so later in other life situations outside of mathematics classes. That is, we submit that the immediacy and mundaneness of the problems may cause students to believe in the applicability of mathematics.

★ Having the formal mathematics grounded in the situation of a fiber art aids the utility of the formal mathematics. It is useful in analyzing the particular fiber art. In addition, the grounding seems to help in transfer to other situations by lending language accessible to the students.

4 Current and Future Research

The mathematics that arises in fiber arts such as knitting, crocheting, cross-stitch, and quilting is wide-ranging and includes topology, graph theory, number theory, geometry, and algebra. One possible framework for approaching this intersection of disciplines involves considering the following questions:

★ What sorts of mathematical objects can be constructed using a particular type of fiber art?

★ What sorts of mathematical concepts can be illustrated using a particular type of fiber art?

★ What intrinsic mathematics is present in a given fiber art?

★ What problems arise in fiber arts that can be answered using mathematics?

Let's examine these questions in turn.

What sorts of mathematical objects can be constructed using a particular type of fiber art? Every fiber art has features and limitations that respectively suggest and restrict the types of objects appropriate for that art. For example, knitting is done with but one strand of yarn. So, one can make a torus by knitting, but not the standard "linear" embedding of a three-holed torus—unless one "cheats" by sewing pieces together. Really, the question is asking what can be made in a mathematically meaningful way using a puristic form of a particular fiber art.

With knitting, we can construct topological surfaces (see Chapter 4 for a project and [3] for more general theory), flat polygons and circles (Amy Szczepański is thinking about uniformly positively curved surfaces), and pieces of uniformly negatively curved surfaces (see Chapter 10). The Möbius band is fairly common, but the obvious way to do it (knit a strip, give it a half-twist, and graft) is not mathematically satisfying. Miles Reid [20] and Maria Iano seem to have been the first to design the intrinsic-twist Möbius band pattern most of us use (albeit in modified forms).

Crochet is significantly different from knitting in that one can build up a third dimension while still using a single strand of yarn. Thus, objects like the Sierpinski triangle (see Chapter 3) and space-filling curves (explored by Miyuki Kawamura) can be crocheted. With crochet one can also create hyperbolic surfaces (see [15]) and orientable topological surfaces. However, it seems that nonorientable surfaces are not possible because such surfaces have only one side and thus require a reversible stitch (which does not yet exist in the crochet lexicon).

Any tiling of the plane may be made using quilting. In fact, tilings of other surfaces, such as the Möbius band (see Chapter 1) or the hyperbolic plane (explored by Daina Taimina) are possible as well. Using sewing more generally, we can make a wide variety of objects, but it is difficult to mimic uniform curvature. Chapter 7 examines a historical model of the projective plane.

It would be interesting to see future work that not only exhibits new mathematical objects we can make, but also lists objects that *cannot* be made with various fiber arts. The latter question is, of course, more difficult, because we must provide proof that the objects cannot be constructed.

What sorts of mathematical concepts can be illustrated using a particular type of fiber art? This has some overlap with the previous question. After all, constructing a projective plane illustrates the concept of nonorientability! But, in asking specifically about concepts, this question encompasses ideas and patterns (which are not objects) as well. Interestingly, the differences between the fiber arts make them more or less suitable to illustrating particular concepts.

Temari balls (see [25]) provide excellent demonstrations of duality for spherical polyhedra. While one could knit or crochet such polyhedra, it seems artificial to force a surface that is naturally flexible to mimic the rigidity of polyhedral faces; temari balls are already spherical.

Because crochet makes such a firm fabric, one can examine the geometric aspects of a crocheted surface such as the hyperbolic plane (see [15]). For example, one can mark geodesics (straight lines relative to the surface) and make measurements on the surface. Daina Taimina has constructed a proof, in crochet, that the helicoid and catenoid are the same object with different parametrizations.

Many fiber arts exhibit symmetry patterns in their design, and in particular frieze and wallpaper patterns are popular. However, cross-stitch is particularly suited to exhibiting them (see Chapter 5) because it is always done on a flat surface that resembles a plane.

Cabling is a common knitting element in sweaters that closely mimics a mathematical structure known as a braid group (see Chapter 8). Intarsia knitting (large patches of color) corresponds to creating a map on the knitted object and dually corresponds to drawing a vertex-colored graph on a knitted object. Therefore, many concepts in topological graph theory may be visualized via knitting.

Quilting is the perfect medium for proofs-without-words, which are visual (usually geometric) representations of proofs; the only one we've seen is of the Pythagorean Theorem (explored by Anabeth Dollins and Janice Ewing). There is lots of room for future work here! Group multiplication tables (explored by Gwen Fisher) and group structures such as cosets (explored by Amy Szczepański) are nicely illustrated by quilts and their topstitching. This is another example where the practices of a particular fiber art make it more conducive to showing certain concepts; topstitching allows underlying commonality to be shown across certain areas of a quilt without interfering with the viewer's perception of the overall structure on the quilt design.

Of course, the research project in this direction should be to find more examples of concepts nicely illustrated by fiber arts.

What intrinsic mathematics is present in a given fiber art? That is, what mathematics arises because of the practices or structure of a fiber art?

The mathematical theory of knots contains three ways of untangling yarn, known as the Reidemeister moves. Both knitting and crocheting are done via a sequence of Type II Reidemeister moves. Dan Isaksen is working on a theory of cohomology (higher derived limits) in knitting. Blackwork embroidery uses graph traversals (see Chapter 9).

Thinking about fiber arts in this way is quite new, and not much has been done with it; in fact, it's not clear whether there is a great deal of potential or very little.

What problems arise in fiber arts that can be answered using mathematics? We are considering pure mathematics and so exclude engineering mathematics of the textile industries, like the dynamics of torque in spun yarn or thread.

Much of knitting is an exercise in arithmetic; for example, computing the gauge of a swatch and using it to size the various parts of a piece of clothing uses nothing more than addition and multiplication. Yet even problems consisting of only simple mathematics are quite complicated until they are correctly stated, and the correct statement is also the job of the mathematician. Nonetheless, more advanced mathematics appears as well, in picking up stitches (see Chapter 2), using cyclic groups while making socks (see Chapter 6), designing reversible fabric (see [2]), and even computing the curve of a neckline with calculus (sarah-marie actually did this while at a party).

Creating sewn garments or objects requires finding separations of surfaces into local patches, where the coordinatization and metric matter. That's a difficult problem in differential geometry that seamstresses regularly solve by eye. Resizing garments is mathematically challenging, because more ideas than the obvious proportionality are involved (for example, does one want to scale the area, the perimeter, or … ?). An exercise in Mary Harris' *Cabbage* was to combine two different sizes of (old, worn-out) bags into one larger-size (new) bag. Placing pattern pieces on fabric is really an optimization problem, and sometimes involves restrictions induced by the grain of the fabric, patterns on the fabric that must be matched along seam allowances, and so forth.

Temari balls involve frighteningly deep mathematics if one wishes to create them precisely. One major issue is plotting points on the sphere, since even the circumference is unknown. In reality, the circumference is not constant, which complicates matters, but even dealing with an ideal sphere, it is complex to plot the vertices of, say, an icosahedron with the traditional method that uses a piece of paper tape. (Note that we have access only to the exterior of the sphere.) Additional problems include studying polyhedra with vertices on the sphere, other tilings of the sphere, and all of the problems of embroidery. Of course, one can also acknowledge the fact that the ball is not exactly spherical, which opens up all kinds of other problems. In addition, the paper tape can be folded rather than marked with a ruler, which brings in origami methods. Further, rather than precise point markings, we may allow approximate markings. This again widens our scope of both questions and techniques.

Like knitting, crochet involves a large amount of counting. However, advanced mathematics can inform a wide array of techniques. For example, lace tablecloths sometimes consist of pieces all of one shape; sometimes they consist of pieces of two different shapes, because the plane cannot be tiled by just one of the shapes. Also, when making structures known in crochet as pineapples, the shape of the points created relies on factors such as the initial number of decrease loops and lengths of the chains making up these loops. Using mathematics can help to determine how to choose these values to create a desired shape. The other way to choose such values is by trial and error. A completely different kind of question is, when can crochet be used to capture the beautiful iron trellis work of Italy and New Orleans? This question is primarily mathematical in that the main difference between crochet and trellis work is the necessity of crochet to be connected in a number of places in order to maintain structural integrity, whereas trellis work may have multiple freestanding curlicues. As is evident, the variety of questions is vast.

Embroidery, cross-stitch, and needlepoint have some common mathematical problems. One is to determine both possible and optimal paths for the needle to take through the cloth to complete a given pattern. Another is minimizing thread usage. These are dealt with in [4], work of Barbara Ashton and Kevin Dove, and our Chapter 9; graph theory plays a prominent role.

There are surely other places where mathematics can illuminate issues in fiber arts, and we look forward to seeing them arise.

5   Conclusion

Over the course of putting together this book, we have carried around the chapter projects to airports, conferences, meetings, and parties. At each venue people have asked excitedly, "What's that? Is it hard to make? What does it mean?" These conversations always involve the questions "When is the book coming out? Where can I get it?" and "Will I be able to understand it?" You already know the first two of these three questions. To the third the answer is both "Yes" and "Hopefully, not quite." That is, there is something in here you can absolutely understand, and we hope that there are things in the book that you will want to learn and can learn.

A final question is always, "How did you think of putting these ideas together?" As avid fiber artists, who have to actively avoid people who might introduce us to new hobbies; as trained mathematicians and lovers of mathematics; as teachers, who think deeply about how students learn and about how we ourselves learn; and as people who cannot compartmentalize their lives but intellectualize their experiences for maximum enjoyment, we have been discussing ideas like these for years.

Now, our fervent desire is that if you picked up the book for the patterns, you will try reading the Overview sections to see how the math feels. If you are reading this book for the math, we hope that you will try out the teaching ideas to hook your students further into mathematics. And, if you got this book to help make your classroom a little more snazzy, we hope you will try to make some of the projects yourself. The interplay between the first three sections of each chapter is amazing, and you will be surprised how much more fulfilling each part is when combined with the others.

Again, welcome—we hope you enjoy the book. Go read the rest of it.

Bibliography

All Fiber Arts except Weaving

[1] Adams, Colin, Fleming, Thomas, and Koegel, Christopher. "Brunnian Clothes on the Runway: Not for the Bashful." *American Mathematical Monthly*, vol. 111, no. 9, November 2004, pp. 741–748.

[2] belcastro, sarah-marie and Yackel, Carolyn. "About Knitting." *Math Horizons*, vol. 14, November 2006, pp. 24–27, 39.

[3] belcastro, sarah-marie. *Knitting Topological Surfaces*. In preparation.

[4] Biedl, Therese, Horton, John D., and Lopez-Ortiz, Alejandro. "Cross-Stitching Using Little Thread." In *Proceedings of the 17th Canadian Conference on Computational Geometry (CCCG'05)*, pp. 199–202. Available at http://www.cccg.ca/proceedings/2005/54.pdf.

[5] Cochrane, Paul. "Knitting Maths." *Mathematics Teaching*, September 1988, pp. 26–28.

[6] Conway, J. H. "Mrs. Perkins's Quilt." *Mathematical Proceedings of the Cambridge Philosophical Society*, vol. 60, 1964, pp. 363–368.

[7] DeTemple, Duane. "Reflection Borders for Patchwork Quilts." *Mathematics Teacher*, February 1986, pp. 138–143.

[8] Fisher, Gwen. "Quilt Designs Using Non-Edge-to-Edge Tilings by Squares." In *Meeting Alhambra, ISAMA-BRIDGES Conference Proceedings 2003*, edited by R. Sarhangi and C. Sequin, pp. 265–272. University of Granada Publication, Granada, 2003.

[9] Fisher, Gwen. "The Quaternions Quilts." *FOCUS*, vol. 25, no. 1, 2005, pp. 4–5.

[10] Fisher, Gwen and Medina, Elsa. "Cayley Tables as Quilt Designs." In *Meeting Alhambra, ISAMA-BRIDGES Conference Proceedings 2003*, edited by R. Sarhangi and C. Sequin, pp. 553–554. University of Granada Publication, Granada, 2003.

[11] Funahashi, Tatsushi, Yamada, Masashi, Seki, Hirohisa, and Itoh, Hidenori. "A Technique for Representing Cloth Shapes and Generating 3-Dimensional Knitting Shapes." *Forma*, vol. 14, no. 3, 1999, pp. 239–248.

[12] Griffin, Mary. "Wear Your Own Theory!: A Beginner's Guide to Random Knitting." *New Scientist*, March 26, 1987, pp. 69–70.

[13] Harris, Mary. "Mathematics and Fabrics." *Mathematics Teaching*, vol. 120, 1987, pp. 43–45.

[14] Harris, Mary. *Common Threads: Women, Mathematics and Work*. Trentham Books, Stoke-on-Trent, 1997.

[15] Henderson, David W. and Taimina, Daina. "Crocheting the Hyperbolic Plane." *The Mathematical Intelligencer*, vol. 23, no. 2, 2001, pp. 17–28.

[16] Irving, Claire. "Making the Real Projective Plane." *Mathematical Gazette*, November 2005, pp. 417–423.

[17] Isaksen, Daniel and Petrofsky, Al. "Möbius Knitting." In *Bridges: Mathematical Connections in Art, Music, and Science*, edited by R. Sarhangi, pp. 67–76. Tarquin, St. Albans, 1999.

[18] Mabbs, Louise. "Fabric Sculpture—Jacob's Ladder." In *Bridges London, BRIDGES Conference Proceedings 2006*, edited by R. Sarhangi and J. Sharp, pp. 561–568. Tarquin, St. Albans, 2006.

[19] Osinga, Hinke M. and Krauskopf, Bernd. "Crocheting the Lorenz Manifold." *The Mathematical Intelligencer*, vol. 26, no. 4, 2004, pp. 25–37.

[20] Reid, Miles. "The Knitting of Surfaces." *Eureka—The Journal of the Archimedeans*, vol. 34, 1971, pp. 21–26.

[21] Ross, Joan. "How to Make a Möbius Hat by Crocheting." *Mathematics Teacher*, vol. 78, 1985, pp. 268–269.

[22] Trustrum, G. B. "Mrs. Perkins's Quilt." *Mathematical Proceedings of the Cambridge Philosophical Society*, vol. 61, 1965, pp. 7–11.

[23] Williams, Mary C. "Quilts Inspired by Mathematics." In *Meeting Alhambra, ISAMA-BRIDGES Conference Proceedings 2003*, edited by R. Sarhangi and C. Sequin, pp. 393–399. Tarquin, St. Albans, 2003.

[24] Williams, Mary C. and Sharp, John. "A Collaborative Parabolic Quilt." In *Bridges: Mathematical Connections in Art, Music, and Science, Conference Proceedings 2002*, edited by R. Sarhangi, pp. 143–149. Tarquin, St. Albans, 2002.

[25] Yackel, C. A. "Embroidering Polyhedra on Temari Balls." In *Math+Art=X Boulder, CO Conference Proceedings 2005*, pp. 183–187, 2005.

Weaving

[26] Clapham, C. R. J. "When a Fabric Hangs Together." *Bulletin of the London Mathematical Society*, vol. 12, no. 3, 1980, pp. 161–164.

[27] Clapham, C. R. J. "The Bipartite Tournament Associated with a Fabric." *Discrete Mathematics*, vol. 57, no. 1–2, 1985, pp. 195–197.

[28] Clapham, C. R. J. "When a Three-Way Fabric Hangs Together." *Journal of Combinatorial Theory Series B*, vol. 38, no. 2, 1985, p. 190.

[29] Clapham, C. R. J. "The Strength of a Fabric." *Bulletin of the London Mathematical Society*, vol. 26, no. 2, 1994, pp. 127–131.

[30] Grünbaum, B. and Shephard, G. C. "Satins and Twills: Introduction to the Geometry of Fabrics." *Mathematics Magazine*, vol. 53, no. 3, 1980, pp. 139–161.

[31] Grünbaum, B. and Shephard, G. C. "A Catalogue of Isonemal Fabrics." *Discrete Geometry and Convexity, Annals of the New York Academy of Sciences*, vol. 440, 1985, pp. 279–298.

[32] Grünbaum, B. and Shephard, G. C. "An Extension to the Catalogue of Isonemal Fabrics." *Discrete Mathematics*, vol. 60 (1986), 155–192.

[33] Grünbaum, B. and Shephard, G. C. "Isonemal Fabrics." *American Mathematical Monthly*, vol. 95, 1988, pp. 5–30.

[34] Hoskins, J. A. "Factoring Binary Matrices: A Weaver's Approach." In *Combinatorial Mathematics, IX (Brisbane, 1981)*, pp. 300–326, Lecture Notes in Mathematics 952. Springer, Berlin-New York, 1982.

[35] Hoskins, J. A., "Binary Interlacement Arrays and Structural Cross-Sections." *Congressus Numerantium*, vol. 40, 1983, pp. 63–76.

[36] Hoskins, Janet A. and Hoskins, W. D. "The Solution of Certain Matrix Equations Arising from the Structural Analysis of Woven Fabrics." *Ars Combinatoria*, vol. 11, 1981, pp. 51–59.

[37] Hoskins, J. A. and Hoskins, W. D. "An Algorithm for Color Factoring a Matrix." In *Current Trends in Matrix Theory (Auburn, AL, 1986)*, pp. 147–154. North-Holland, New York, 1987.

[38] Hoskins, Janet A., Praeger, Cheryl E., and Street, Anne Penfold. "Balanced Twills with Bounded Float Length." *Congressus Numerantium*, vol. 40, 1983, pp. 77–89.

[39] Hoskins, Janet A., Praeger, Cheryl E., and Street, Anne Penfold. "Twills with Bounded Float Length. *Bulletin of the Australian Mathematical Society*, vol. 28, no. 2, 1983, pp. 255–281.

[40] Hoskins, J. A., Hoskins, W. D., Street, Anne Penfold, and Stanton, R. G. "Some Elementary Isonemal Binary Matrices." *Ars Combinatoria*, vol. 13, 1982, pp. 3–38.

[41] Hoskins, J. A., Stanton, R. G., and Street, Anne Penfold. "Enumerating the Compound Twillins." *Congressus Numerantium*, vol. 38, 1983, pp. 3–22.

[42] Hoskins, J. A., Stanton, R. G., and Street, A. P. "The Compound Twillins: Reflection at an Element." *Ars Combinatoria*, vol. 17, 1984, pp. 177–190.

[43] Hoskins, Janet A., Street, Anne Penfold, and Stanton, R. G. "Binary Interlacement Arrays, and How to Find Them." *Congressus Numerantium*, vol. 42, 1984, pp. 321–376.

[44] Hoskins, J. A. and Thomas, R. S. D. "The Patterns of the Isonemal Two-Colour Two-Way Two-Fold Fabrics." *Bulletin of the Australian Mathematical Society*, vol. 44, no. 1, 1991, pp. 33–43.

[45] Lucas, E. *Application de l'Arithmétique à la Construction de l'Armure des Satins Réguliers*. Paris, 1867.

[46] Lucas, E. "Principii fondamentali della geometria dei tessute." *L'Ingegneria Civile e le Arti Industriali*, vol. 6, 1880, pp. 104–111, 113–115.

[47] Lucas, E. "Les principes fondamentaux de la géometrie des tissus." *Compte Rendu de L'Association Française pour l'Avancement des Sciences*, vol. 40, 1911, pp. 72–88.

[48] Pedersen, Jean J. "Some Isonemal Fabrics on Polyhedral Surfaces." In *The Geometric Vein*, pp. 99–122. Springer, New York-Berlin, 1981.

[49] Pedersen, Jean. "Geometry: The Unity of Theory and Practice." *The Mathematical Intelligencer*, vol. 5, no. 4, 1983, pp. 37–49.

[50] Roth, Richard L. "The Symmetry Groups of Periodic Isonemal Fabrics." *Geometriae Dedicata*, vol. 48, 1993, pp. 191–210.

[51] Roth, Richard L. "Perfect Colorings of Isonemal Fabrics Using Two Colors." *Geometriae Dedicata*, vol. 56, 1995, pp. 307–326.

[52] Shorter, S. A. "The Mathematical Theory of the Sateen Arrangement." *The Mathematical Gazette*, vol. 10, 1920, pp. 92–97.

[53] Thomas, R. S. D. "Isonemal Prefabrics with Only Parallel Axes of Symmetry." Preprint, December 2006.

[54] Woods, H. J. "Part II—Nets and Sateens, of the Geometrical Basis of Pattern Design." *Textile Institute of Manchester Journal*, vol. 26, 1935, pp. T293–T308.

Mathematics Education

[55] Freudenthal, Hans. *Revisiting Mathematics Education: China Lectures*. Kluwer, Dordrecht, 1991.

[56] Gravemeijer, Koeno Pay Eskelhoff. *Developing Realistic Mathematics Education*. Freudenthal Institute, Utrecht, 1994.

[57] Mason, John. *The Discipline of Noticing, Sunrise Research Laboratory*. RMIT, Melbourne, 1992.

[58] Polya, G. *How to Solve It*, Second Edition. Princeton University Press, Princeton, NJ, 1971.

[59] Steffe, L. and D'Ambrosio, B. "Toward a Working Model of Constructivist Teaching: A Reaction to Simon." *Journal for Research in Mathematics Education*, vol. 26, no. 2, 1995, pp. 146–159.

CHAPTER 1

quilted möbius band

AMY F. SZCZEPAŃSKI

1 Overview

The world of quilting is rich in applications of mathematics. Quilters rely on geometry when designing and constructing their quilts, so examples abound of primary and secondary school lesson plans that connect the geometry found in traditional American patchwork quilts to school geometry (see Section 3). Our discussion will be restricted to quilts based on traditional American patchwork; contemporary quilts and art quilts often use a variety of other techniques and materials. The Möbius quilts described in this chapter are somewhat non-traditional in appearance, as they are not flat, but they are constructed using fairly traditional methods.

For those not familiar with quilting, the main ingredients of a quilt are fabric, batting, and thread. The fabric is typically cotton printed with a design, and most of the fabrics used have a "right" side and a "wrong" side, just as clothing is "right side out" or "inside out." An unstated assumption of traditional quilting is that only the right side of the fabric may show. Wrong sides, seams, and unfinished edges are always hidden.

When constructing a quilt, the quilter cuts out patches of fabric and sews them together to form what is known as the *quilt top*. This is the colorful, geometric arrangement of patches that normally comes to mind when one thinks about quilts. Once the top is completed, it is layered over the *batting* (the fluffy inner layer that provides insulation) and a large piece of fabric known as the *backing* (also called the *lining*). The right sides of the top and the backing face out; the wrong sides face the batting. These three layers are held together by decorative stitching known as *quilting*. Once the quilting is completed, the quilt is trimmed to size, and the edges are covered with *binding*. For more information on the techniques of quilting, see [7].

Typically geometry works behind the scenes in service of a quilt design. The quilter may use geometry to calculate the amount of fabric to purchase or to arrange the patches and blocks in a symmetrical design. Geometry is also used in deciding how to cut out triangular patches so that they will be the right size after being sewn together with a $\frac{1}{4}''$ seam allowance. Applications of symmetry and tiling to traditional quilt blocks and their extensions can be found in [2]. A more artistically based exploration of the uses of geometry is in [11]; this book by an MIT alumna and quilt-maker includes her thoughts on symmetry, tessellations, and fractals as inspirations for arranging quilt patches. In contrast, the quilt projects in this chapter were designed to showcase the features of a shape known as the Möbius strip or Möbius band (see Figure 1). Another project involving the Möbius band appears in Chapter 7.

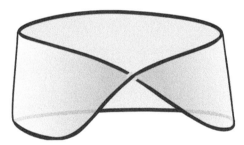

Figure 1. A basic illustration of a Möbius band.

Constructing a quilted Möbius band is similar to constructing a paper Möbius band. A strip of quilt is assembled, twisted, and the ends seamed together; binding is then used to hide the raw edges.

The Möbius band has appeared in other fabric arts. In the mid-1930s fashion designer Madeleine Vionnet (credited with the invention of the bias cut) designed dresses and coats that featured scarves shaped like Möbius bands [9].

1.1 Experiments with a Möbius Band

If you've never seen a Möbius band before, you should make a model of one now. Cut a strip approximately 11″ long and 1″ wide from a sheet of paper. (Accuracy is not important for this; use whatever paper you have handy.) Bring the short sides of the paper together as if you were going to form a loop, but before joining the edges flip one of the ends over, introducing a twist of 180°. Now tape the ends together. You have just made a Möbius band! See Figure 2 for a photograph of the completed shape.

Figure 2. Paper Möbius band.

These six experiments will illustrate some of the most unexpected properties of the Möbius band and will generate paper models for the quilts in Section 4. They are also meant to provide inspiration for designing your own variations on the Möbius quilt. Before trying each experiment, attempt to use your geometric intuition to anticipate the result.

These first two experiments illustrate that the Möbius band has only one side.

Experiment 1 Using a paper Möbius band and a pen, draw a line lengthwise down the center of the strip. See Figure 3 for how to begin drawing the line. Continue until your line meets up with itself.

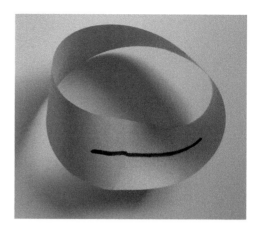

Figure 3. The start of a line drawn down the middle of a Möbius band.

Experiment 2 With a second paper Möbius band and a pen, draw a line lengthwise about one third of the way over from the left edge. See Figure 4 for how to begin drawing the line. Continue until your line meets up with itself.

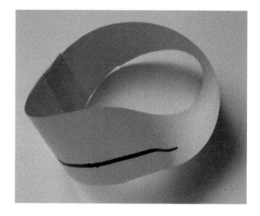

Figure 4. The start of a line drawn one third of the way from the edge.

The lines you drew with the pen should have ended where they began; while they may have started on what seems like the "front" of the strip, they continued onto the "back"—showing that there is, in fact, only one side to a Möbius band. Because there is no separate front and back, any lengthwise line drawn on the surface will traverse the entire strip. Notice that the second line appears both near the left and the right of the sides of the strip.

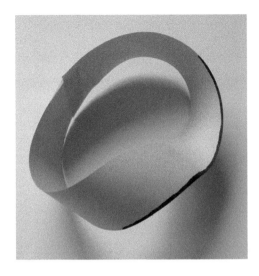

Figure 5. The start of a line drawn along the edge.

Experiment 3 With a third paper Möbius band and a magic marker (or some other wet-ink pen), draw a line along the very edge of the strip, allowing the ink to soak into the paper. See Figure 5.

The third experiment shows that the Möbius band has only one edge: the "left" edge and the "right" edge are the same.

Experiment 4 Using the strips from Experiments 1 and 2, cut along the lines that you drew. (You may need to pinch the paper to begin cutting.)

During this experiment, the strip with the line down the middle should have become a longer, thinner strip with more twists (but still in one piece). The strip with the line one-third of the way over should have divided into two linked loops of paper: one should look like a thinner version of the original Möbius strip, while the other should be longer and have more twists. This phenomenon is explained in Section 2.1.

Experiment 5 Take two strips of paper, each about 1″ by 11″, and hold them together as a double thickness. Bend this double-thick strip into the shape of a Möbius band. Using two pieces of tape (one for each "side"),

tape together the ends where they meet. Shake out the strip and spread it out the best you can. How many sides does your shape have? How many twists are in it?

The result of Experiment 5 is a loop of paper with four half-twists, that is, a total twist of $720°$.

You can carry out this experiment in reverse by taking a long, narrow piece of paper, introducing four half-twists, attaching the edges, and coaxing the resulting shape into a double-thickness Möbius strip. It takes some patience to get this to work out, so don't get discouraged if it seems difficult at first.

The sixth experiment comes from [1] and shows that the Möbius strip reverses orientations.

Experiment 6 Using a paper Möbius strip and a hole punch with an asymmetrical design, punch holes along the length of the strip as in Figure 6. If you don't have a such a hole punch, you can make your Möbius strip out of tracing paper and draw an asymmetrical design with a dark pen. What happens when you return to the starting point?

Figure 6. Starting to punch spiral holes.

Once the pattern has been continued along the entire strip, there should be mirror image versions of the design adjacent to each other.

Further experiments can be found in [1] and [15]. The results of all of these experiments can be explained by the mathematics of the Möbius strip, which is introduced in Section 2.

Figure 7. A quilt where left hands turn into right hands.

1.2 Extending the Experiments to Fabric

The properties of the Möbius band shown in the paper models also inform the design process for a Möbius band made out of fabric. In terms of this project, the most important property is that the Möbius band has only one side! While most fabric has a "right" side and a "wrong" side, the Möbius band does not. Because traditional quilt construction endeavors to hide the wrong side of the fabric, quilting is an ideal medium for the creation of textile Möbius bands. Were we to make this shape by taking our paper experiments and repeating them with unquilted fabric, we would end up with products with right sides sewn to wrong sides and where the seam and raw edges showed.

The one-sided nature of the Möbius band is important to consider when choosing a pattern for quilting together the layers of a Möbius strip quilt. If you choose a quilting pattern with a row of stitching 2″ from the right edge of the strip and continue sewing until you reach your starting point again, your line of stitches will—like the line drawn in Experiment 2—travel along the strip twice, giving you a line 2″ from the right edge *and* a line 2″ from the left edge.

A property that mathematicians call nonorientability influences the quilting pattern as well. This property implies, roughly, that if a Möbius band was quilted with an asymmetrical pattern, such as left hands (see Figure 7), that at some point along the shape the left hands would appear to switch to right hands. Because of this, any asymmetric quilting pattern will eventually meet up with its reverse. You can choose to highlight this feature or not. To showcase the reversal of orientation, repeat an asymmetrical quilting pattern or cut-out; to hide it, choose a symmetrical pattern or be sure to include both versions of an asymmetrical pattern along the length of the strip.

Like a traditional quilt, a Möbius quilt has only one edge to bind. However, unlike a traditional quilt, the binding will not need to go around any corners and will follow a smooth, continuous path. Experiment 4 shows what happens during the process of binding a quilted Möbius band. When trimming the edges prior to binding a traditional quilt, the scraps will simply fall off. However, with a Möbius quilt, the trimmed edge becomes a twisted loop wrapped around the main project, and the scrap will need to be snipped crosswise in order to be removed.

Experiment 4 forms the basis of one of the quilt patterns in Section 4.1; a pattern is given for a Möbius quilt with a zipper down the middle. A Möbius band cut one third of the way from the edge can also be illustrated in fabric by the use of zippers or other fasteners. There is an unexpected difficulty in a zippered quilt, arising not from the construction but from an aspect of its display: Once the quilt is unzipped, how should the resulting loop be twisted to allow the quilt to be re-zipped? You may want to try this with a cut paper Möbius band: Can you twist the loop of paper around to return the Möbius band to its original shape? This takes some practice and some trial and error; sometimes a little bit of decorative embroidery can be used to help remember how to re-zip the quilt.

When unzipped, the Möbius quilt in Section 4.1 is the same shape as the strip of paper created in Experiment 5. This suggests an alternate, elegant method for constructing a quilted Möbius band with fewer seams than the typical construction of creating a strip of quilt, introducing a half twist, and sewing the ends together. Running this experiment in reverse forms the basis for the pattern in Section 4.2. There, four half-twists are made in a long strip of fabric, which is then seamed in one place and formed into a double-thickness Möbius band; batting is inserted between the layers, and the edges are bound. Because this Möbius band is of double thickness, the right side of the fabric is all that shows.

2   Mathematics

2.1   Explaining the Experiments

An effective way to visualize what happens in the experiments is to introduce the idea of a polygonal representation of a surface. This idea is defined with great precision in [14, Chapter 12]; this construction and its application to surfaces such as the Möbius band and the Klein bottle (see Chapter 7) is also described in [10, Chapter 1]. It can be generally described as giving directions on how to start with a polygonal region and attach its edges in order to create the desired shape. These instructions specify which edges are to be glued together and which way they are to be aligned (roughly, twisted or untwisted).

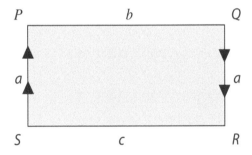

Figure 8. A polygonal representation of the Möbius band.

A polygonal representation for the Möbius band is illustrated in Figure 8. Sides labeled with the same letter are attached to each other so that the directions of the arrows match. In this example, the side labeled a is glued, twisted, to the other side labeled a. As a result, corner P is attached to corner R, and corner Q is attached to corner S. In some ways, this polygonal representation resembles a video game where characters who travel past the right edge of the screen return on the left edge of the screen. If the game world were shaped like a Möbius band, then characters who went off the right side near the top of the screen would return on the left side near the bottom of the screen.

Looking at the Möbius strip this way, we can see that it has only one edge. If we travel along the edge PQ of the polygonal representation from left to right, when we reach point Q, we continue at the point S to which it is connected. From there, continuing from left to right, we will eventually reach point R, which is connected to P, bringing us back where we started. What had been two edges in the original polygon is now just one edge in the Möbius strip. To emphasize the way that the edges are identified, we often give attached corners the same letters.

We can use this representation of the Möbius band to explain what happens when we cut the band down the middle, as is illustrated in Figure 9. Because our original Möbius band has only one edge, our cut down the middle will create a second edge. This explains why the result of our experiment is a two-edged loop. After cutting, this loop is still in one piece because the edges labeled a remain attached, as do the edges labeled d. Cutting along the center line crosses the original identified edge twice, resulting in two half twists.

Similar analyses explain the other experiments.

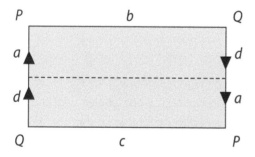

Figure 9. Using a polygonal representation to explain what happens when the Möbius band is cut down the middle.

2.2   The Origin of the Möbius Band

The Möbius band was discovered in 1858 by the German mathematician August Ferdinand Möbius and independently a few months earlier by Johann Benedict Listing. Möbius was both an astronomer and a mathematician,

and his work focused on topology, symmetry, and celestial mechanics [6, p. 121] and extended at times to actuarial science, acoustics, and other applications of the mathematical sciences [6, p. 19].

One of Möbius' original uses for his surface was to introduce the concept now called nonorientability. He described the band in terms of polygons being pieced together [6, p. 108]; the shape can be visualized as triangular patches with rules about how the edges are attached (or seamed). Depending what rules one follows when attaching the edges, one may obtain a cylinder, a Möbius band, or a different twisted shape. Möbius compared his one-sided band to two-sided shapes by describing the implications of assigning a "handedness" to the triangles. With our familiar two-sided surfaces, we can define "clockwise" for the triangles in a consistent way that respects the seams between the triangles; in the Möbius band, however, this cannot be done.

So far in this chapter the word "geometry" has been used to categorize the mathematics under consideration, but a more accurate term would be "topology." Möbius' construction of his nonorientable surface provided a key ingredient in the solution of the topological problem of classifying surfaces. The other ingredient is the Euler characteristic, denoted χ. The *Euler characteristic* counts the number of holes in a surface that can be made of polygons glued together at their edges. For a solid with v vertices, e edges, and f faces, it is calculated as $v - e + f$; this quantity depends on the number of holes in the solid and will equal 2 if it has no holes. See [16, Section 11.2] for more information on the Euler characteristic.

Theorem 1   [16] *Two polygonal representations define homeomorphic surfaces if and only if they have the same Euler characteristic and the same orientability character.*

In the eighteenth century, the Swiss mathematician Simon-Antoine-Jean Lhuilier was working on the problem of classifying solids where $v - e + f \neq 2$

[6, p. 106]. Completing the classification requires the "orientability character," that is, a determination of whether a surface is one-sided or two-sided. Möbius developed a method to precisely describe this idea of one- and two-sidedness for any surface formed from pieced polygons. His method was to assign the same orientation to every polygon that formed the shape; they would all be assigned either "clockwise" or "counterclockwise." He required that polygons that share an edge have compatible orientations along that edge. Orientations are *compatible* if they point in opposite directions along the edge [6, p. 106]. A surface is said to be *orientable* or two-sided if it is possible to assign compatible orientations to all the polygons that are glued together to form it. If this is impossible, then the surface is *nonorientable* or one-sided.

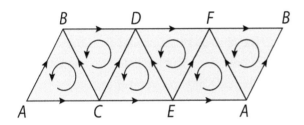

Figure 10. Cylinder *Y* with compatible orientations.

Figure 10 illustrates a surface with a compatible orientation. Notice that this is a polygonal representation of a cylinder, with the two copies of edge *AB* identified without a twist. All the triangles are given counterclockwise orientations. Consider edge *BC*. In triangle *ABC*, its orientation points from *C* to *B*; in triangle *BCD*, this edge is oriented from *B* to *C*. Because the orientation points in different directions along the shared edge, the orientation is compatible. This can be seen for all the other triangles with shared edges in the figure, including the glued edge.

Alternatively, Figure 11 gives a polygonal representation of a Möbius band with an incompatible orientation. The only difference between Figure 11 and

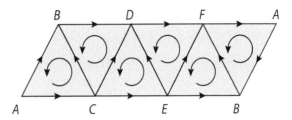

Figure 11. Möbius band *M* with incompatible orientations.

Figure 10 is the direction in which the two copies of edge *AB* are attached. What had previously been a compatible orientation is now incompatible, as the joining edge is connected with a twist.

While the Möbius band was first developed as a theoretical idea to illustrate the limitations of "handedness" for some figures created by joining polygons, it has achieved widespread popularity. Both [6] and [15] give examples of its use in art, music, and literature. Practical uses, including twisted resistors and twisted conveyor belts that don't wear out as quickly as conventional ones, are also described in [15].

2.3 Homology

The criterion given above is not useful for showing that a surface is nonorientable: How do we show that there cannot exist a compatible orientation system for the surface-subdividing polygons? After all, perhaps we just chose polygons badly, and some other set of polygons would have a compatible orientation system.

Luckily, homology provides an answer to our dilemma. For one thing, homology calculations are independent of the subdivision of the surface into polygons (see Chapter 2 of [13]). Furthermore, we can reformulate Möbius's criterion in a decidable form using homology: we make a certain calculation, and depending on its value we know whether the surface is orientable or nonorientable.

Recall that to calculate simplicial homology for a Δ-complex such as shown in Figures 10 and 11, we begin by forming a chain complex

$$\cdots C_3 \xrightarrow{\partial_3} C_2 \xrightarrow{\partial_2} C_1 \xrightarrow{\partial_1} C_0 \xrightarrow{\partial_0} 0$$

where C_n is the free abelian group with generators the n-dimensional simplices of our object and ∂_n is the boundary map defined by summing the $(n-1)$-dimensional components of an element of C_n with signs determined by the orientations on the n-simplices and $(n-1)$-simplices.

Thus, in both Figure 10 and Figure 11, we have the chain complex

$$\cdots 0 \xrightarrow{\partial_3} \mathbb{Z}^6 \xrightarrow{\partial_2} \mathbb{Z}^{12} \xrightarrow{\partial_1} \mathbb{Z}^6 \xrightarrow{\partial_0} 0$$

as each object has 6 vertices, 12 edges, and 6 triangles.

The *nth homology group* H_n is defined as $\ker \partial_n / \operatorname{im} \partial_{n+1}$. As examples, we compute H_1 of the cylinder Y and Möbius band M. We label all edges and 2-simplices in alphabetical order of their vertices.

By noting that

$$\operatorname{im} \partial_1(Y) = \operatorname{im} \partial_1(M)$$
$$= \langle B - A, C - A, D - B, E - C, F - D \rangle$$

so that $\ker \partial_1$ must have seven generators, we see that

$$\ker \partial_1(Y) = \langle AC + BC - AB, CD - BD - BC,$$
$$CE + DE - CD, EF - DF - DE,$$
$$AE + AF - EF, AB - BF - AF,$$
$$AC + CE + AE \rangle$$

($BD + DF + BF$ is equal to the sum of those seven generators) and

$$\ker \partial_1(M) = \langle AC + BC - AB, CD - BD - BC,$$
$$CE + DE - CD, EF - DF - DE,$$
$$BE + BF - EF, -AB - BF - AF,$$
$$AC + CE + BE + BD + DF + AF \rangle.$$

Then

$$\operatorname{im} \partial_1(Y) = \langle AC + BC - AB, CD - BD - BC, CE + DE - CD,$$
$$EF - DF - DE, AE + AF - EF, AB - BF - AF \rangle$$

and

$$\operatorname{im} \partial_1(M) = \langle AC + BC - AB, CD - BD - BC, CE + DE - CD,$$
$$EF - DF - DE, BE + BF - EF, -AB - BF - AF \rangle.$$

Thus,

$$H_1(Y, \mathbb{Z}) = \langle AC + CD + AE \rangle \cong \mathbb{Z},$$

and

$$H_1(M, \mathbb{Z}) = \langle AC + CE + BE + BD + DF + AF \rangle \cong \mathbb{Z}.$$

Now we may describe the usual homological criterion for orientability: A connected surface without boundary components is orientable if and only if $H_2(M, \mathbb{Z}) \neq 0$. Although this is an easy criterion, it doesn't apply to the Möbius strip or cylinder, as these two surfaces have boundary. To formulate the orientability criterion in this case, it is convenient to consider relative homology.

Recall that if $R \subset S$ is an inclusion of Δ-complexes, then the relative chain groups C_n are the quotient groups $C_n(S)/C_n(R)$, and the boundary operator ∂_n on $C_n(S)$ induces a boundary operator on the chain complex formed by the relative C_n groups. The *relative homology* $H_n(S, R; \mathbb{Z})$ is then defined as $\ker \partial_n / \operatorname{im} \partial_{n+1}$ as before. We can now state a more general criterion for orientability:

A connected surface S with boundary ∂S is orientable if and only if $H_2(S, \partial S; \mathbb{Z}) \neq 0$. As $\partial_3 = 0$ for a surface, this implies that S is orientable if and only if the chain map $\partial_2 \colon C_2(S) \to C_1(S)/C_1(\partial S)$ has nontrivial kernel. (Note that $C_2(\partial S)$ is omitted because it is trivial.)

The group $C_2(Y)$ is spanned by the 2-simplices ABC, BCD, CDE, DEF, AEF, and ABF, and $C_1(Y)/C_1(\partial Y)$ is spanned by the 1-simplices AB, BC, CD, DE, EF, and AF. Writing ∂_2 in matrix form for the cylinder Y we get

	ABC	BCD	CDE	DEF	AEF	ABF
AB	−1	0	0	0	0	1
BC	1	−1	0	0	0	0
CD	0	1	−1	0	0	0
DE	0	0	1	−1	0	0
EF	0	0	0	1	−1	0
AF	0	0	0	0	1	−1

This happens to be a square matrix, so to check if it is singular, one need only calculate the determinant. In this case, the determinant is 0, so ker $\partial_2 \neq 0$, and the cylinder is orientable.

The group $C_2(M)$ is spanned by the 2-simplices ABC, BCD, CDE, DEF, BEF, and ABF, and $C_1(M)/C_1(\partial M)$ is spanned by the 1-simplices AB, BC, CD, DE, EF, and BF. The ∂_2 matrix for M differs from that for Y only in the upper right-hand entry:

	ABC	BCD	CDE	DEF	BEF	ABF
AB	−1	0	0	0	0	−1
BC	1	−1	0	0	0	0
CD	0	1	−1	0	0	0
DE	0	0	1	−1	0	0
EF	0	0	0	1	−1	0
BF	0	0	0	0	1	−1

In this case, the determinant is 2, implying that $H_2(M, \partial M; \mathbb{Z}) = \ker \partial_2 = 0$. Thus the Möbius strip is nonorientable.

This formulation of orientability encompasses Möbius' original idea, because any orientation set on a polygonal decomposition of a surface corresponds to assigning signs in the boundary maps. These define a consistent orientation if and only if the signed sum of the polygons represents a nontrivial relative homology class.

Some exercises that may be interesting for graduate students to consider are found in Section 3.2.

2.4   Graphs on Möbius Bands

Surfaces vary in the number of colors necessary to vertex-color graphs on them. While the four-color the-orem is the most well-known result of this sort, there are versions of this theorem for surfaces other than the sphere (and, equivalently, the plane). An upper bound based on the Euler characteristic χ of the surface is known as the Heawood bound $h(\chi)$. Equality holds for every closed surface with $\chi \leq 1$ except the Klein bottle.

Theorem 2   [3] *The chromatic number of a graph G drawn on a closed surface of Euler characteristic* $\chi \leq 1$ *is at most* $h(\chi) = \lfloor (7 + \sqrt{49 - 24\chi})/2 \rfloor$.

In particular, any graph on the Klein bottle ($\chi = 0$) or on the projective plane ($\chi = 1$) can be colored with at most six colors [3, p. 157]. While this theorem is stated in terms of closed surfaces, we can apply it to the Möbius band by examining these nonorientable surfaces; when a disk is removed from a projective plane, what remains is a Möbius strip. The question comes down to determining the largest complete graph that can be drawn on the surface with no edges crossing. Since the Möbius band is contained in both the Klein bottle and the projective plane, any graph on a Möbius band can be drawn on a Klein bottle or a projective plane. Therefore, a graph on a Möbius band will require at most six colors as well. As the complete graph on six vertices can be drawn on the Möbius band, there exists a map on the Möbius band requiring six colors.

A graph coloring problem can be stated as a map coloring problem by considering the map that is dual to the graph. Countries represent vertices, and borders between countries represent edges. Considering map colorings on the Möbius band leads to some nice classroom applications, as we will see in Section 3.3.

3   Teaching Ideas

Quilting has been used to motivate a wide variety of classroom activities. One of the most common involves symmetry and tessellation [8], but quilt-based lessons can incorporate ideas as diverse as modular arithmetic, transformational geometry, and number operations [5]. Because many regions and cultures have their own quilting traditions, a quilt-based math lesson may also serve as a jumping-off point for interdisciplinary study [12].

Activities involving Möbius bands can be used with students at every level, from elementary school to college. These topics are particularly well-suited for Liberal Arts Math, courses for pre-service elementary and middle school teachers, and geometry/topology courses for pre-service secondary math teachers. Graduate students might tackle the questions in Section 3.2. With younger students these activities can be used any time there is a reason to supplement the standard curriculum. For most students, the lesson can start with building a paper model of the Möbius band and exploring its unusual properties, carrying out experiments similar to those in Section 1.1. In classes below the math-major level, there are usually students who have never encountered the Möbius band before.

The quilted projects described in this chapter, especially the Möbius band with the zipper, can be used as display models to demonstrate the actions that students should be taking with their own paper models of the strips. Additionally, students enrolled in an inter-disciplinary or project-based course may be encouraged to enhance their understanding of the Möbius band by making one of the projects in Section 4 or a crafted Möbius band of their own design.

3.1 Experiments and Conjectures

Students can build their geometric intuition, form conjectures, and then experimentally verify them by carrying out explorations based on the experiments in Section 1.1. Those who need structure to complete a lesson successfully might be asked to go through a scripted set of experiments, respond to specific prompts, and report on their findings. Others who can work more independently may be asked to formulate their own conjectures and to perform the experiments necessary to justify them. More activities of this sort can be found in [1].

As the Möbius band was first developed in order to distinguish between types of surfaces, a nice way to introduce students to the Möbius band is in that setting. Following the treatment in [4, Section 5.2], students can be asked to compare the features of a Möbius band to those of an untwisted loop.

Each student will need several strips of paper approximately $1'' \times 11''$, something to write with, and tape and scissors. While it is possible to share the tape and scissors, it works most efficiently when each student has his or her own supplies.

Before having the students form Möbius bands and perform tasks like those outlined in Experiments 1–6, first instruct them to make loops by taping the ends of the paper together without a twist. Have the students observe the number of sides, the number of edges, and what happens if the loop is cut lengthwise. (This provides a "control group" or point of comparison.) Once they have completed this task, tell them how to form Möbius bands and ask them to perform similar experiments. If time permits, ask the students more questions: What happens with different numbers of twists? What happens with different cuts? What happens if you cut a shape more than once (Experiment 6)? Encourage the students to guess what will happen before carrying out each experiment. When concluding the lesson, ensure that the students know what properties make the Möbius band special and how it differs from other, more familiar shapes.

Another question that students can consider begins by imagining a quilted Möbius band with a longitudinal zipper not in the center, as if to illustrate Experiment 2 in Section 1.1. Suppose that you wanted to construct this quilt so that each of the resulting pieces (the Möbius strip and the twisted loop linked to it) requires the same amount of fabric. Where would you locate the zipper?

(Answer: The zipper should be one fourth of the way from the edge.) Can you design a quilting pattern that illustrates a specified property of the Möbius band?

In addition to these activities based on inductive reasoning, students with the ability to understand the mathematics at work can be asked to deductively explain why the results hold.

3.2 Advanced Questions on Homology

Graduate students learning about homology can deepen their understanding by developing quilt designs that topologically illustrate the algebraic calculations. Can you design a quilt that exhibits, either in the fabric choice or in the topstitching, the fact that $H_1(M, \mathbb{Z}) = \mathbb{Z}$? Can you design a pair of quilts that exhibit, either in the fabric choice or in the topstitching, the fact that Y is orientable but M is nonorientable according to the homology definition of orientability?

3.3 Map-Coloring Problems: How to Arrange Quilt Patches

Questions about arranging colored patches on a quilted shape are closely related to map-coloring problems—often referred to in the literature as graph-coloring problems (see Section 2.4). In these problems, the goal is to color-code a map so that no two adjacent regions are the same color. Shapes that meet at a corner, such as diagonally adjacent squares on a checkerboard, are allowed to be the same color. We then ask how many colors are necessary to color maps that satisfy specified conditions. Because of the Four Color Theorem, it is known that a map in the plane (equivalently, the sphere) can be colored with at most four colors. This is tantamount to stating that any patchwork quilt with adjacent patches made of different fabrics can be sewn using at most four different fabrics.

In 1840, August Möbius posed the following puzzle:

There was once a king with five sons. In his will he stated that after his death the sons should divide the kingdom into five regions in such a way that the boundary of each one should have a frontier line in common with each of the other four. Can the terms of the will be satisfied? [6, p. 14]

The conditions of Möbius' problem cannot be satisfied if the kingdom is on the surface of the Earth. Students can consider this problem on other surfaces.

Ask the students to show that the terms of the king's will can be met on a Möbius strip. Since a Möbius strip is only one-sided, for the sake of this exercise, you should assume that it is transparent. To make a physical model of the situation, each student will need strips of overhead transparency, tape, and markers. Students with more advanced spatial skills may be able to complete the task without such aids. Once the students discover a way to divvy up the land so that each brother's land borders all of the others' (equivalently, a quilt design on a Möbius strip that requires five different fabrics), challenge the students to try for six. Examples of maps requiring five and six colors, respectively, are illustrated in Figures 12 and 13. Once students have successfully created maps needing five or six colors, challenge them to refine their maps. Can they make maps where the boundaries are all straight lines? Can they make a map with symmetrical patches? Can they make a map where the patches are congruent?

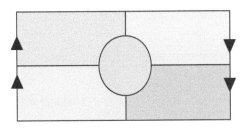

Figure 12. A map that requires five colors.

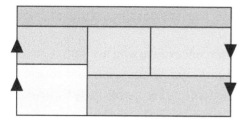

Figure 13. A map that requires six colors.

Students may also be asked to describe the construction process of a Möbius quilt that illustrates the 5- or 6-color map. Because the Möbius strip should be functionally transparent, how should the fabric patches be cut out and sewn together so that the quilt looks the same from either "side"? (See Figure 14.)

Figure 14. A map needing six colors on a Möbius quilt.

For related map-coloring exercises (including a sample quilt), see Chapter 7.

4 How to Make Möbius Quilts

These patterns assume that the reader is familiar with the basic techniques of sewing. Unless otherwise stated, all seam allowances are $\frac{1}{4}''$.

4.1 A Möbius Band with a Zipper

These instructions produce a large model of the Möbius band that has been cut down the middle, as in Exper-

iment 3 in Section 1.1. The zipper allows for the band to be separated and reattached at will. When unzipping the completed model, take care to note how the fabric twists and lines up. Sometimes it takes a little practice to learn how to zip it back together.

Materials

* $1\frac{1}{2}$ yards of 42″ wide cotton quilting fabric (or $\frac{3}{4}$ yards of 60″ wide fabric).

* Two pieces of quilt batting, each approximately 7″ by 50″.

* 3 yards of single-fold binding, either bias or straight-grain. Binding strips should be approximately $2\frac{1}{4}''$ wide before folding or $1\frac{1}{8}''$ wide after folding.

* 48″ separating zipper, such as the Parka Dual-Separating Zipper (article F44) by Coats and Clark.

* Thread for construction of the quilt and for quilting.

Instructions

1. Cut out four rectangles of fabric, each measuring $50\frac{1}{2}''$ long and 6″ wide. The length of the rectangle should be the length of your zipper tape plus 1″. It's better to err on the side of too long than too short. The little extra length can be trimmed later.

2. Separate the zipper and baste (either by machine, using a zipper foot, or by hand) each half to the right side of the long edge of one of the rectangles with the teeth pointing toward the center of the fabric and the edge of the zipper tape aligned with the raw edge. Center the zipper along the length of the rectangle.

3. Make two piles as follows: The bottom layer will be a rectangle of fabric with the zipper basted to it, laying right side up on the work surface. On top of that, place a piece of fabric without the zipper, wrong side up. These two pieces of fabric should have right sides together, and their raw edges should be aligned. On top of the fabric, place a piece of batting. The batting

Figure 15. The zippered quilt.

is larger than the fabric. Align the edge of the batting with the long edge of fabric that has the zipper; allow the other edges of the batting to overhang the fabric by about an inch on each side.

4. Flip the piles over so that you can see the basting where you sewed the zipper. For each pile, sew the long edge with the zipper, using a zipper foot (if desired) and being careful not to break your needle by stitching over the zipper pull. The stitching should pass through both pieces of fabric, the zipper tape, and the batting. Remove the basting if it shows on the right side.

5. Trim the batting from the seam allowances to reduce bulk.

6. Turn each pile right side out. Press as needed.

7. Use the zipper to attach your two pieces to form one strip.

8. Trim the open short edges, leaving $\frac{1}{2}''$ seam allowance beyond the end of zipper tapes. You will be sewing two seams to attach the edges. One will attach the top half of the left edge to the bottom half of the right edge, and the other will connect the top half of the right edge to the bottom half of the left edge.

9. Twist into a Möbius shape. With the right sides together and abutting each edge, fold and pin through the fabric layer of the top on one side and of the bottom on other side. Unzip 6 '' from each end; open out batting and fabric and align each cut edge with right sides together. Sew $\frac{1}{4}''$ seam through the tops, catching the batting in the seam. Begin your seam at the outside edge of the Möbius band, continue past the zipper, and end at the outside edge. Trim batting; press into a continuous strip. Do the same for the other set of short edges.

10. Quilt the Möbius band with the quilting pattern of your choice.

Figure 16. The quilt, unzipped.

11. Trim the remaining raw edge of the shape so that the edges of the fabric and the batting are aligned.

12. Apply the binding to cover the raw edge, using the method of your choice. See [7] for a complete discussion about how to bind a quilt. I prefer an all-machine method of application. Align the raw edge of the binding with the raw edge of the quilt, leaving several inches of binding free at the beginning. Sew the binding to the quilt. When you return to the starting point, join the ends of the binding as neatly as possible. Fold the binding over to cover the raw edge so that the folded edge aligns with your previous stitch line. Using a narrow zig-zag stitch, sew the folded edge in place.

13. When first unzipping the completed quilt, take note of how the parts fit together. Reassembling it is not as easy as it might seem. A small decorative motif can serve as a reminder of how it is aligned when zipped back together.

4.2 A Möbius Band Seemingly Missing Some Seams

In contrast to the seams and stitching that hold together the previous pattern, this small wholecloth Möbius band has only one seam along its length. People who neither sew nor understand the Möbius band may not appreciate the simple elegance of this project.

Materials

★ 2 yards of 42″ wide cotton quilting fabric. This pattern illustrates the mathematics best when made from a single strip of fabric. To use less yardage, you can make the strip both shorter and narrower. In that case, you could use $\frac{1}{4}$ yard of 60″ wide fabric.

★ A piece of quilt batting approximately $7\frac{1}{2}''$ wide and 38″ long. If you choose to alter this pattern to use less fabric, the batting should be the same width as your cut rectangle of fabric (no overhang) and a little bit longer than half as long as the rectangle.

Figure 17. Seamless Möbius.

* $4\frac{1}{4}$ yards of single-fold bias or straight grain binding. Binding strips should be approximately $2\frac{1}{4}''$ wide before folding or $1\frac{1}{8}''$ wide after folding.

* Thread for construction of the quilt and for quilting.

Instructions

1. Start by making a paper model of the project: take a long, narrow strip of paper, twist it 720°, tape the edges together, and finagle it into a double-thickness Möbius band. (This is Experiment 5 from Section 1.1.)

2. Cut a piece of fabric $7\frac{1}{2}''$ wide and $70\frac{1}{2}''$ long.

3. Put four half-twists (total rotation of 720°) into the fabric. It doesn't matter whether you twist clockwise or counterclockwise. Sew the short edges together, right sides together. Press the seam open or to one side. (See Figure 18.)

Figure 18. Four half twists.

4. Following your paper model, manipulate the fabric into a double-thickness Möbius band with the right side facing out. Pull the fabric taut. It is possible to arrange the fabric so that the shape is either right-side-out or wrong-side-out [15, p. 10]. If your shape is

wrong-side-out, you do not need to rip out the seam; you just need to rearrange the fabric.

5. Carefully insert the piece of batting between the layers. Tug on the fabric as necessary to assure that the fabric is evenly distributed around the strip. Align the long edges of the batting with the raw edges of the fabric.

6. Once the batting is aligned, trim it to length, abut the short edges, and baste them together by hand. Realign the batting with the fabric.

7. Quilt the Möbius band, if desired. To keep the layers from shifting, pin them together with safety pins.

8. Bind the raw edges with the binding strips. See the instructions for the Möbius band with a zipper for more information on binding.

Bibliography

[1] Barr, Stephen. *Experiments in Topology*. Crowell, New York, 1964.

[2] Beyer, Jinny. *Designing Tessellations: The Secrets of Interlocking Patterns*. Contemporary Books, Chicago, 1999.

[3] Bollobás, Béla. *Modern Graph Theory*. Springer-Verlag, New York, 1998.

[4] Burger, Edward B., and Starbird, Michael. *The Heart of Mathematics: An Invitation to Effective Thinking*, Second Edition. Key College Publishing, Emeryville, CA, 2005.

[5] Ernie, Kathryn T. "Mathematics and Quilting." In *1995 Yearbook: Connecting Mathematics across the Curriculum*, edited by Peggy A. House and Arthur F. Coxford, pp. 170–176. National Council of Teachers of Mathematics, Reston, VA, 1995.

[6] Fauvel, John, Flood, Raymond, and Wilson, Robin (eds.). *Möbius and His Band: Mathematics and Astronomy in Nineteenth-Century Germany*. Oxford University Press, New York, 1993.

[7] Fons, Marianne, and Porter, Liz. *Quilter's Complete Guide*. Oxmoor House, Birmingham, AL, 1996.

[8] Garrett Anthony, Holly, and Hackenberg, Amy J. "Making Quilts without Sewing: Investigating Planar Symmetries in Southern Quilts." *Mathematics Teacher*, vol. 99, no. 4, November 2005, p. 270.

[9] Kirke, Betty. *Madeleine Vionnet* (edited by Harumi Tokai). Kyuryudo Art Publishing Co., Ltd., Tokyo, 1991.

[10] Massey, W. S. *Algebraic Topology: An Introduction*. Springer-Verlag, New York, 1967.

[11] McDowell, Ruth B. *Art & Inspirations: Ruth B. McDowell*. C&T Publishing, Lafayette, CA, 1996.

[12] Moyer, Patricia S. "Patterns and Symmetry: Reflections of Culture." *Teaching Children Mathematics*, vol. 8, no. 3, November 2001, pp. 140–144.

[13] Munkres, James R. *Elements of Algebraic Topology*. Addison-Wesley, Reading, MA, 1984.

[14] Munkres, James R. *Topology*, Second Edition. Prentice Hall, Upper Saddle River, NJ, 2000.

[15] Pickover, Clifford A. *The Möbius Strip: Dr. August Möbius's Marvelous Band in Mathematics, Games, Literature, Art, Technology, and Cosmology*. Thunder's Mouth Press, New York, 2006.

[16] Stahl, Saul. *Geometry from Euclid to Knots*. Pearson Education, Upper Saddle River, NJ, 2003.

CHAPTER 2

picking up stitches and diophantine equations

LANA HOLDEN

1   Overview

Many a knitter's secret shame is a bag, perhaps stashed in the back of a closet or under a bed, of knitted pieces that languish just a few hours' work short of becoming a functional piece of knitwear. This phenomenon is due to a common aversion to an aspect of the knitting process known as "finishing." Finishing consists of those activities that must be performed after most of the knitting (i.e., the "fun part") is completed, in order to transform a pile of knitted fabric into a useful (and, one hopes, attractive) object.

A distaste for sewing together various pieces of knitted fabric also provokes many knitters to limit themselves to items that can be created as a single piece. The simplest of these is the traditional rectangular scarf or wrap, which is knit back and forth in rows until the knitter either (a) decides it's long enough, (b) tires of the project, or (c) runs out of yarn. Regardless, when knitting ceases, the knitter needs only to weave in yarn tails—no sewing is necessary. Hats can be knitted in the round, that is, in a continuous spiral, in order to avoid seams. Mittens and gloves are little more than very small hats where some of the stitches secede into separate spirals partway from wrist to fingertip. Traditional sock construction is a hybrid process where leg and foot are worked in the round but the heel is worked in rows. Even several traditional sweater styles can be knit seamlessly in the round (see [2, Chapter 4] for the seminal treatment).

All of these items have something in common: they are worked in a single direction. But what if we want two sections of knitting to meet perpendicularly without sewing? If we are clever, we can avoid the use of the sewing needle by use of the technique of *picking up stitches*, whereby a new section of knitting is begun along the side edge of an already-completed section.

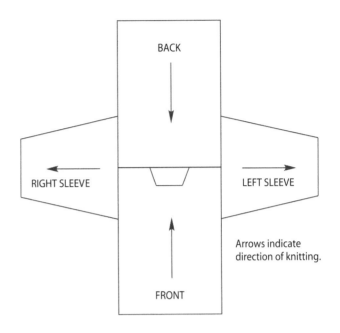

Figure 1. Schematic of a drop-shouldered sweater.

Figure 2. A top-down sleeve in progress.

A straightforward instance of picking up stitches occurs when the knitter has completed the front and back of a drop-shouldered sweater (see Figure 1), has joined the shoulders (perhaps avoiding sewing with *three-needle bindoff* [1, p. 75]), and chooses to knit the sleeves directly onto the body from the shoulder down as in Figure 2.

This chapter's project (see Section 4) is a hat in which the body is knitted onto the side edge of the brim by picking up stitches.

1.1 How to Pick up Stitches

The mathematician will probably be satisfied to be assured that a method exists for creating a new row of knit stitches between two columns of stitches, so that work can proceed in a perpendicular direction. However, the knitter wishes to know how to execute the method. These instructions are given for right-handed knitters; the left-handed knitter should refer to Chapter 4 for help in translating the instructions.

1. Hold your work with the right side facing you and the pickup edge at the top. Working from right to left, insert the right hand needle from front to back between the first and second stitch of the desired row. (See Figure 3.)

Figure 3. Inserting the needle between the first two stitches of the row.

2. Wrap the working yarn around the needle in your usual manner. Carefully bring the loop of yarn through to the front of the work. (See Figure 4.)

Figure 4. Making a picked-up stitch.

Repeat steps 1 and 2 from right to left along the pickup edge.

Knitters who find this maneuver challenging may use a crochet hook to draw the loop of yarn to the front of the work, and then transfer the loop to the knitting needle.

Note that although Figures 3 and 4 illustrate picking up stitches from the knit side of stockinette, stitches may be picked up from the purl side instead, or along the edge of another stitch pattern entirely. In any case, the needle is inserted from the intended right side of the garment, one stitch in from the edge. The hat in Section 4 (see also Figure 5) requires picking up stitches from the purl side as in Figure 9, where it can be useful to look over the top of the work to be sure the needle pokes through the knit side in the proper place.

Figure 5. A variant of the hat in a self-striping yarn with very long color shifts.

1.2 Where to Pick up Stitches

For each row end along the selvedge of knitted fabric, the knitter may choose to either pick up (create) a stitch between the first two stitches of that row, or skip that row end. At this point, some arithmetic presents itself.

The ratio of stitches per inch to rows per inch is generally less than 1 : 1 and depends on the stitch pattern being used in the sweater. For the most commonly used stitch, known as stockinette stitch (see Figure 6), the ratio of stitches to rows is approximately 5 : 6.

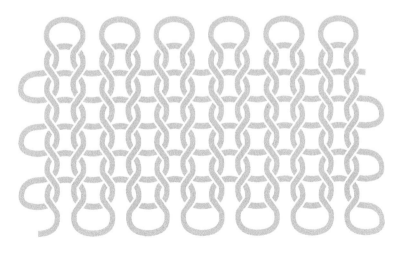

Figure 6. Structure of stockinette stitch.

If we want our sweater to lie smoothly, we cannot simply pick up a stitch at the end of each row, or we will have too many stitches, and the sleeve will puff all around the armhole. Knitting instructions commonly read something like, "Pick up 82 stitches evenly spaced between armhole markers." But perhaps there are 96 rows, and hence 96 possible opportunities to pick up a stitch, between those markers. How do we decide which row ends to use to pick up stitches and which to skip?

In practice, the computation-phobic knitter will probably eyeball this and get satisfactory results with a little trial and error. A fussier and more arithmetic-friendly knitter will at least calculate that the ratio of 82 to 96 is in the ballpark of 5 to 6, and hence skip one out of every six rows with a little fudging at the beginning and end; this usually produces very nice results (and a finished sweater) without the application of any number theory whatsoever.

The mathematically minded knitter, however, asks, "How do we define, precisely, what it means for picked-up stitches to be 'evenly spaced'? And, further, is there an algorithm for doing so?" The concept of even spacing can be defined as a system of equations; it turns out that a solution to this system provides a useful decision method for stitch placement. We now describe this method, which is derived in Section 2.

Algorithm 1 (Which stitches to pick up.) Suppose we are told to pick up s stitches along the selvedge of a completed section of knitting with r rows.

1. Calculate $R = (r - s) + 1$.

2. Using division with remainder, compute s/R to obtain the quotient, ℓ, and the remainder, b.

3. Define $a = R - b$.

This tells us that we should pick up ℓ stitches and then skip a row end a times, and pick up $\ell + 1$ stitches and then skip a row end b times. In practice, deciding

how to interleave the sequences of ℓ stitches with the sequences of $\ell + 1$ stitches is easy. How best to arrange these picked-up stitches depends on the values of a and b; we discuss this problem in Section 2.3.

The algorithm just presented is used in the bi-directional hat project of Section 4 and is the source of educational activities of Section 3. In the following section, we provide all the mathematical details behind the algorithm.

2 Mathematics

In this section, we will model the problem of evenly spacing picked up stitches using a set of Diophantine equations, comment on the existence and uniqueness of solutions to these equations, and describe an open problem in implementing the solutions. First, we need to describe sequences of picked-up stitches rigorously.

Definition 1 A *run* is a set of consecutive row ends in which stitches are picked up and which is preceded and followed by one row end in which no stitch is picked up. The *length* of a run is the number of stitches picked up in that run. An *ℓ-run* is a run of length ℓ.

2.1 Modeling the Problem

Let us examine what we wish it to mean for picked-up stitches to be spaced evenly. Motivated by both structural and aesthetic considerations, we desire as much consistency as possible.

★ We require that neither the first row nor the last row be skipped, since we want the new section of knitting to span the full length of the chosen previous section.

★ We require that no two consecutive rows be skipped, lest we create a visible hole or pucker in the knitting.

* We restrict the lengths of runs to at most two adjacent integers (e.g., we might only pick up runs of three stitches or four stitches).

We might also desire, from an ease-of-knitting perspective, that the runs be as long as possible; however, we will show later that there is only one possible pair of lengths satisfying the previous three constraints.

We may now set up our system of equations. Notice that because stitches, rows, and runs are tangible and discrete, all of our variables represent non-negative integers. Thus, our system will be Diophantine. Let r be the number of row ends and s be the number of stitches to pick up. We will consider only cases where $r \geq s$, since this is true in practice. Let ℓ represent the smaller of the two run lengths required by the third constraint, so that some runs have length $\ell + 1$. We need to determine the run lengths and the number of runs of each length, i.e., we need to find ℓ and non-negative integers a and b such that

$$a(\ell) + b(\ell + 1) = s. \tag{1}$$

This equation represents the statement that all the stitches in runs of both lengths combined add up to the total number of picked-up stitches.

Consider the number of row ends: it must equal all the picked-up stitches plus the skipped row ends. Exactly one skipped row end follows each run except the last one, by the first two constraints. So we have

$$a(\ell) + b(\ell + 1) + a + b - 1 = r. \tag{2}$$

2.2 Existence and Uniqueness of Solutions

It is not immediately apparent from the equations above that any solutions exist. (How do we know we haven't made life too hard for ourselves?) If, however, we subtract Equation (1) from Equation (2), we can observe that the number of runs is exactly one more than the number of skipped row ends:

$$a + b = (r - s) + 1. \tag{3}$$

If we rearrange Equation (1), we get

$$\ell(a + b) + b = s,$$

and substituting for $a + b$ using Equation (3) yields

$$\ell((r - s) + 1) + b = s.$$

Recall that $r \geq s$, hence $r - s + 1 > 0$. So we can find ℓ and b using the Division Algorithm (to non-mathematicians, this is ordinary division), and then $a = ((r - s) + 1) - b$.

Now that we know at least one solution (ℓ, a, b) exists, the next obvious question is whether there is more than one solution. We did not specify that $b < r - s + 1$, i.e, that the remainder be less than the dividend, so we cannot simply cite the Division Algorithm to claim uniqueness. However, from Equation (3) we have $b = r - s + 1 - a$, so as long as $a \neq 0$, we do have $b < r - s + 1$; and because a was defined to be the number of runs of length ℓ, of which there is always at least one, we do have $a \neq 0$.

2.3 Additional Questions

Although we now know how to find the number of runs of each length to pick up, we must still decide how to arrange the runs. We want to maximize symmetry and to interperse the two run lengths evenly along the edge; it is not immediately clear how those concepts should be defined rigorously. Here are a few examples of how a knitter might choose to arrange the runs in rather obvious instances:

* To pick up a runs of length ℓ and $a + 1$ runs of length $\ell + 1$: Begin by picking up an ℓ-length run and alternate run lengths, ending with an $(\ell + 1)$-length run.

* To pick up 1 run of length ℓ and a runs of length $\ell + 1$ where a is even: Pick up the ℓ-run in the center, with $a/2$ $(\ell + 1)$-runs on either side.

* To pick up 2 runs of length ℓ and a runs of length $\ell + 1$ where a is even: Begin by picking up an ℓ-length run, then pick up $a/2$ $(\ell + 1)$-runs, then the other ℓ-length run, then pick up the remaining $a/2$ $(\ell + 1)$-runs.

The situation already becomes mathematically murky in the relatively simple case of picking up 1 ℓ-run and a $(\ell + 1)$-runs where a is odd. We can't center the lone ℓ-run exactly. (In practice, a knitter would probably place it just off-center.) In the following section, this problem becomes a possibility for classroom exploration.

3   Teaching Ideas

A question commonly posed by students of all levels is, "Why do I need to know how to do this by hand? I have a calculator!" Picking up stitches evenly illustrates an occurrence of long division where the whole number remainder is essential information; a decimal remainder generated by a calculator is insufficient. See Algorithm 1, which can be used in responding to the common student complaint or as a practical application for the Division Algorithm in a number theory class.

For students learning ratios, picking up stitches along swatches of a variety of stitch patterns can provide a visual connection to the concept of ratios. For example, students might be provided with samples of stockinette, garter, and seed stitch. Students may then be asked to determine experimentally what ratio of stitches to rows provides the most even change of direction.

The symmetry issues discussed in Section 2.3 can serve as open-ended questions for pattern exploration.

As an introduction to group theory, or just to the lower-level concept of divisibility, a teacher might ask students to investigate the case of picking up stitches around a circle. Here is a sample investigation:

Suppose you need to pick up stitches around a circular band with 36 rows.

* Could you pick up two stitches and then skip a row end all the way around? How many stitches would you pick up all together?

* What about picking up three stitches and then skipping a row? four stitches? five stitches?

* How can you tell which numbers will work out evenly?

* What might you do if you need to pick up 29 stitches? What is interesting about the number 29? (Answer: 29 is prime.)

4   How to Make a Bi-Directional Hat

The adult's rolled-brim beanie shown in Figure 7 uses bi-directional knitting to showcase a self-striping yarn. The mathematics of spacing picked-up stitches evenly is used when knitting the body of the hat onto its brim.

Materials

* 1 ball Noro Silk Garden (just barely enough; it is recommended that you purchase a second ball).

* Size 8 16″ circular needle and double-pointed needles (or size needed to obtain gauge).

Gauge

* 16 stitches and 21 rows = 4″. (Row gauge is not particularly important.)

Figure 7. Photograph of the bi-directional hat.

Figure 8. The hat adjusts easily to yarns of different gauges and is amenable to embellishment.

Figure 9. Picking up stitches from the purl side.

Finished Size

The actual circumference of the hat is 20″. It is designed to fit snugly and comfortably stretches to fit most adults (21″–23″ head circumference). Fit is easily customized by adjusting the length of the brim band (see band directions). Note that the top shaping requires the number of stitches to be evenly divisible by 8; if you choose to alter the size, it is simplest to do so by changing the gauge. Those knitters willing to engage with advanced arithmetic can attempt variants such as those shown in Figure 8.

Instructions

Band of the Hat

The band is worked back and forth in rows, using either two double-pointed needles or a circular needle. Cast on 10 stitches, leaving an 18″ tail for sewing the ends of the band together. (Note: If you are comfortable with a removable cast-on method, you may use it and graft the ends of the band for an apparently seamless join.) Beginning with a purl side row, work in stockinette stitch (purl 1 row, knit 1 row) until band measures 20″ or until band fits snugly when slightly stretched around the head of the intended wearer; end with a knit side row. Using the circular needle, bind off 10 stitches purlwise, leaving the last loop on the needle (if you are planning to graft, purl the ten stitches and place all but the last stitch on a holder). Rotate work a quarter-turn, so that the long edge is at the top and the purl side of the band

is facing you. (You may sew or graft the ends of the band together at any time from now on, or wait until the end.)

Picking Up Stitches

Pick up 80 stitches knitwise (not counting the loop already on the needle) evenly spaced along the long edge of the band, using the technique shown in Figure 9 and detailed below. The number of row ends will vary, depending on each knitter's individual row gauge and size customization; if you have knitted to the size and gauge above, you will have 105 row ends. (Otherwise, refer to Section 1.) In that case, $(r-s)+1 = (105-80)+1 = 26$. Using the Division Algorithm, $80 = (3*26)+2$. So $\ell = 3$ and $b = 2$; hence, $a = 24$ and $\ell + 1 = 4$. This tells us we need to pick up 24 runs of 3 stitches and 2 runs of 4 stitches. To place the two 4-stitch runs on opposite sides of the hat, our pickup instructions would read as follows.

Picking Up Stitches Pattern

(Pick up a stitch in each of next 3 rows, skip 1 row) 6 times; pick up a stitch in each of next 4 rows, skip 1 row; (pick up a stitch in each of next 3 rows, skip 1 row) 12 times; pick up a stitch in each of next 4 rows, skip 1 row; (pick up a stitch in each of next 3 rows, skip 1 row) 6 times, with no row skipped at the end of the sixth time.

Body of the Hat

The body of the hat is worked in the round. Join by passing the first stitch over the last picked-up stitch and place a marker; you now have 80 stitches. Now work stockinette stitch in the round (knit every round) until hat measures 3″ from the first picked-up round.

Top of the Hat

The top of the hat is shaped by dividing the stitches into eight equal groups and decreasing at the end of each group as shown in the pattern for rounds 1–19 below. Change to double-pointed needles when there are too few stitches to fit on the circular needle.

Top of the Hat Pattern

Round 1: (K 8, K2tog) around: 72 sts

Rounds 2 and 3: Knit

Round 4: (K 7, K2tog) around: 64 sts

Round 5 and 6: Knit

Round 7: (K 6, K2tog) around: 56 sts

Rounds 8 and 9: Knit

Round 10: (K 5, K2tog) around: 48 sts

Round 11: Knit

Round 12: (K4, K2tog) around: 40 sts

Round 13: Knit

Round 14: (K3, K2tog) around: 32 sts

Round 15: Knit

Round 16: (K2, K2tog) around: 24 sts

Round 17: Knit

Round 18: (K1, K2tog) around: 16 sts

Round 19: (K2tog) around: 8 sts

Cut yarn with a 12″ tail, draw through remaining stitches and pull snugly. Sew or graft ends of band together, with seam allowance inside the roll. Weave in ends.

Bibliography

[1] Wiseman, Nancie M. *The Knitter's Book of Finishing Techniques.* Martingale & Co., Woodinville, WA, 2002.

[2] Zimmerman, Elizabeth. *Knitting Without Tears.* Simon & Schuster, New York, 1971.

CHAPTER 3

the sierpinski variations: self-similar crochet

D. JACOB WILDSTROM

1 Overview

Simple projects in crochet are often designed to allow arbitrary resizing. Most crochet artisans choose scarves as their first projects because the item can be considered done whenever the artist feels the scarf is long enough. A similar design philosophy can be applied to a shawl, which is a triangle worked from a point rather than a rectangle worked from an edge. Likewise, center-worked projects such as granny squares and certain doilies are worked outward round by round until the size is deemed appropriate.

For most of these projects, the design relies on repetition of a small pattern. For instance, a granny square uses the same stitch pattern on each round, and a shawl can be made with repetitions of a pineapple motif or a fancy stitch. These can create beautiful projects without a large-scale design: they're simply built up by the orderly arrangement of smaller designs. This chapter focuses on large-scale designs that can be resized. With each iteration (repetition) of the design, the object grows significantly.

Large-scale designs have traditionally been the province of the crochet method known as *filet*, where open and solid stitches are arranged into a grid to form pictures. Conventional filet is worked from a design of fixed shape and size, which does not lend itself to iterative development. A filet project is complete when the predetermined picture is done: it can neither plausibly be halted before the design has been depicted nor made significantly larger than this single design. Iterative designs transcend the limitations of both fixed-sized filet designs and simplistic granny squares. They are more complex and can be worked through any number of stages, each of which is aesthetically attractive.

In the second section, we'll formulate more precisely what is meant by "iterative design" as a mathematical concept. Some of the richest iterative designs are fractals and cellular automata. We'll explore why these mathematical objects are such promising potential motifs for iterative design. We'll investigate the Mandelbrot set, the Sierpinski carpet, the Koch snowflake curve, and the Sierpinski triangle. The Sierpinski triangle will be used as a jumping-off point for a discussion of automata, and the process by which an abstract design can be faithfully rendered in crochet.

Intuitively, a design would be simply a picture or a schematic, in which parts of the picture are colored differently to represent either various stitches or assorted colors of yarn. On a large scale, colors could even represent motifs rather than single stitches. To make a design iterative, we want a series of nested designs, one inside the other, so that the process of creation can be halted at a number of stages. This sort of schematic is what, mathematically, we formalize in the next section.

2 Mathematics

In determining which iterative designs are crochetable, three questions arise from our informal construction: First, what do we mean by *design*? Next, what makes a design *iterative*? And lastly, what restrictions on a design would make it *crochetable*?

Crochet fabric forms a surface, and most projects have sections that are approximately flat (where designs are usually exhibited), so we'll be considering a "design" as a coloring of the plane. The plane is conventionally denoted as $\mathbb{R}^2$, and when we speak of a "coloring" of the plane, we really want to associate each point in the plane with a chosen color. In other words, a *coloring* is a function from $\mathbb{R}^2$ to a set of colors (usually denoted by numbers). However, a design isn't just a matter of filling the plane with color. Crochet projects are finite in size, so we want to choose a portion of the plane to consider as the design region. To represent adequately something we can crochet, we need to restrict the design region in two ways. First, as mentioned above, it must be of finite area; second, it must be connected. Formally assembled, we have an explicit definition of design below:

Definition 1 A *design* $D = (C, f, S)$ is a triple consisting of a set of colors C, a function $f : \mathbb{R}^2 \to C$ representing the coloring, and a connected region $S \subset \mathbb{R}^2$ of finite area.

We will be using the set of positive integers $\mathbb{N}$ as our usual color set, so we may write $(\mathbb{N}, f, S)$ as simply (f, S). Notice that f need not be onto, which corresponds to the fact that in our crocheting we will be using only a finite, indeed small, number of colors or stitch types.

It might seem that we could simplify this definition by making f a function on S instead of $\mathbb{R}^2$. After all, why should we care what colors are used in regions of the plane that aren't part of our design? An *iterative* design, however, is one in which we create a design that becomes part of a larger design, and so forth. As we have already chosen colors for the entire plane in the definition of a design, the only complexity added by iteration is the need to pick successively larger regions of the plane for our subsequent designs. We will denote these regions S_1, S_2, and so forth, and define our iterative process as follows:

Definition 2 An *iterative design* $D = (C, f, (S_1, S_2, ...))$ is a triple consisting of a set of colors C, a function $f : \mathbb{R}^2 \to C$ (as in a non-iterative design), and a sequence of connected regions S_i of $\mathbb{R}^2$ of finite area such that $S_i \subset S_{i+1}$ for $i \geq 1$. Given an iterative design D, we refer to the design $D_i = (C, f, S_i)$ as its *ith iteration*.

This is a flexible definition of an iterative design, in spite of our restrictions on the choices of S_i. In fact, it is far too flexible to be of any use in determining what

might be plausibly crocheted, so we introduce two additional conditions:

⋆ *Connectedness.* Assuming we want to work with a single strand of yarn and not stitch several pieces together, we must be able to work each stage of our design in a single piece. Thus, not only must each S_i be a connected region, but we also require everything in S_i that is *not* in S_{i-1} (this region is denoted $S_i \setminus S_{i-1}$) to be connected, since we want to be able to crochet the *new* parts of D_i with a single strand.

⋆ *Granularity.* Assuming we work with the same weight of yarn and size of hook throughout, there is a minimum size that we can render in each color, which is to say, the size of a single stitch or motif. Thus, we require that all single-color regions in our design be no smaller than some minimum size.

Formalizing these criteria, we have, at last, a definition flexible enough to describe a multitude of designs but restrictive enough to admit only designs that might be possible to crochet.

Definition 3 An iterative design (f, S_i) is *crochetable* if the following two conditions hold:

(a) $S_i \setminus S_{i-1}$ is a connected region for all $i > 1$.

(b) There is a bounded, finite-area region $K \subset \mathbb{R}^2$ such that for each point $x \in S_i$ there exists a translation K_x of K such that $x \in K_x \subset S_i$ and K_x is monochromatic.

The second criterion above may seem particularly unintuitive, but in more conventional language, K is a minimum shape for colored regions and design boundaries and serves as the granularity of our design. In other words, K represents a single stitch. In practice, patterns use a square or triangle for K, so that designs drawn on grids with design boundaries along grid lines will necessarily satisfy this condition.

2.1   Fractals as Crochetable Designs

Understanding dual perspectives for viewing a fractal is essential to envisioning how to represent a fractal in crochet. While media images of fractals typically portray an entire fractal, which can be examined more and more closely to reveal more and more detail, another perspective is possible with some fractals. By definition, fractals are self-similar, and some are obtained through an iterative process that can be followed through its discrete stages. These latter fractals can be conceptualized by beginning with the first stage and building up to higher and higher stages. Because only a finite-size stage can be reached, the granularity level can be set to be the smallest color size needed. Of course, just as it is impossible to zoom out from the first stage to view the complete fractal, it is equally impossible to zoom in from a complete fractal to see infinitesimal detail. Therefore, both perspectives are valuable. Nonetheless, because we have a minimum stitch size, we must adopt the iterative perspective and build each fractal by stages.

The primary decision in producing a series of designs is the question of which levels of fractal detail we can introduce at which stages of construction. Fractals whose details are representable in discrete chunks are thus the best ones for our consideration. As a negative example, an approach of increasingly detailed nested designs is wholly inadequate for the well-known Mandelbrot set, pictured in Figure 1. While the Mandelbrot set has self-similarity, starting with the smallest elements and building outward is not feasible because there is no generative approach to the Mandelbrot set that builds it from the center outward. The interior of the Mandelbrot set can be defined as the set of parameters c such that, for $z_0 = 0$ and $z_i = z_{i-1}^2 + c$, it is true that $|z_i| < 2$ for all i. This lends itself to an iterative process in which the nth generation is given by $\{c : |z_n| < 2\}$; however, the first generation of such a process would be the circle $|z| = 2$, and subsequent iterations would

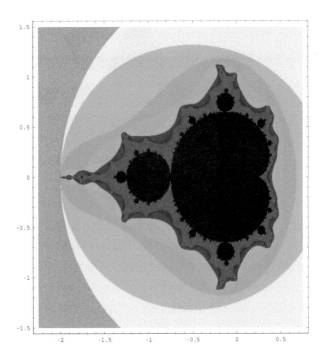

Figure 1. The Mandelbrot set.

be produced by the removal of sections from this design, rather than the addition of new elements on the outside.

The consideration of fractals that are the limits of a generative process allows for a natural correspondence between the concept of an iterative design and a fractal: ideally, we'd like to choose each design region S_i to be identically the ith generation of a fractal.

An example of a fractal that is such a limit of increasingly complicated generations is the Koch snowflake. The first generation of the snowflake is a triangle, and the fractal is built up according to the following rule: at each step, subdivide each edge of the curve into equal thirds, and place an equilateral triangle along the middle third of each side. The first several steps of this process are shown in Figure 2, with the new additions in each generation shown in black.

In setting out to render this as an iterative design, we can start by noticing that the Koch snowflake has only a single color, so our coloring function is trivial and our

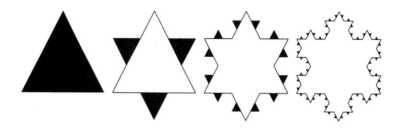

Figure 2. The Koch snowflake, assembled iteratively.

choice of design regions must represent the fractal's increasing complexity. A simple first choice would be to let S_i be the ith generation of the fractal as portrayed in Figure 2, but we may observe that this iterative design is not crochetable. The black regions in the figure represent $S_i \setminus S_{i-1}$, and they are clearly not connected. In addition, the triangles being added become smaller in later iterations, so that detail below the resolution of a single stitch could not be represented at all. As mentioned previously, it is imperative that as we increase detail, we increase the size of our design as well.

The Sierpinski triangle and carpet exhibit a more adaptable form of self-similarity than the other fractals we've seen so far. One traditional construction of the Sierpinski triangle starts with a blue triangle, and pro-

ceeds in each step by subdividing each blue triangle in the fractal into four smaller triangles and coloring each central triangle red. The Sierpinski carpet construction proceeds in much the same manner, but starting with a blue square, and subdividing each blue square into nine smaller squares before coloring the central square red. The first few steps of this process are shown in Figure 3. At first glance this fractal seems ill-suited for iterative crochet—the region in which the design exists is of finite size, and the actual colors within the region change. Furthermore, they do so in a manner that does not appear to obey the granularity condition, as the regions of blue and red become smaller with each iteration. Notice, however, that each iteration is a subset of the following iteration, as shown in Figure 4.

Figure 3. The Sierpinski triangle and carpet, assembled iteratively.

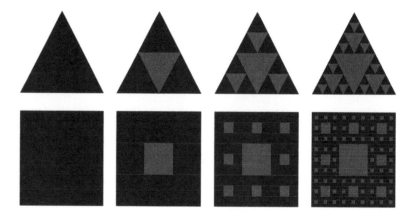

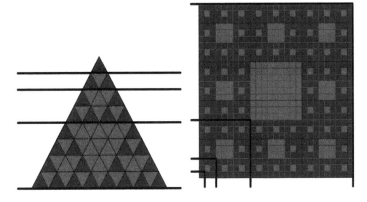

Figure 4. The Sierpinski triangle and carpet, with design subdivisions.

The lines from top to bottom in the triangle and from the bottom-right corner outward in the carpet represent the boundaries of design regions D_1, D_2, and so forth. Each D_i shown in Figure 4 is simply a scaling of the fractal's ith generation as depicted in Figure 3, so this design decomposition does, in fact, allow us to create higher-generation fractals successively as our chosen design region exposes more of the underlying colors.

2.2 Turning Designs into Crochet

We now have two self-similar designs in the abstract sense of a series of nested pictures, but how do we turn this into crochet? Our colors represent choices of

Figure 5. A Sierpinski shawl, colored to emphasize the fractal structure.

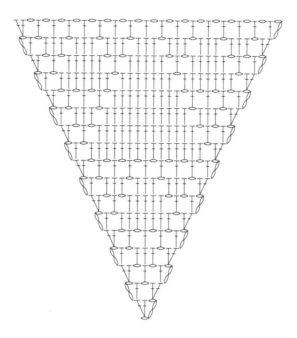

Figure 6. Mary Pat Campbell's design of the Sierpinski triangle.

stitches or motifs (see Figure 5), and, in the end, most as-sembly of stitches and motifs in crochet takes on one of two forms:

1. Back-and-forth work is used for scarves and most other projects with essentially rectangular construc-tion. Here stitches are laid in a row, and at the end of a row the work is turned over.

2. Work in the round is used for hats, doilies, and other projects with radial symmetry. It is performed by lay-ing stitches around the work in concentric rings.

Back-and-forth work is frequently performed as if the stitches formed a rectangular grid: a stitch is roughly rectangular in shape, and while biases are introduced in the ways stitches link together, they cancel out when a piece is worked back-and-forth. When working in the round, the biases produced by stitches' off-center link-age do not cancel out. In patterns worked in the round, certain motifs are arranged with successive rows off-set. Thus our best abstraction of crochet worked in the round is not a rectangular grid, but a system of rectan-gles with successive rows offset by a half-period, as in a brick wall, or a grid of squares viewed at a 45° angle. This model is also representative of back-and-forth cro-chet designs in which successive rows of stitches or mo-tifs are crafted between, rather than directly atop, the motifs or stitches of the previous row.

Mary Pat Campbell first developed such a design for the Sierpinski triangle [1], shown in Figure 6. In or-der to read this diagram, note that the triangle is read from bottom to top and composed of tall, skinny capi-tal Ts with crosses on them and ovals inserted between T tops. The tall Ts represent double crochets and the ovals represent chain stitches. This project has a simple design rule operating throughout the body of the pat-tern, viewing stitches in pairs: in each row, we make two double-crochet stitches if the four stitches of the previ-ous row are all double-crochets or a pair of chain-1 dou-ble crochets, and work a chain-1 followed by a double crochet if the four stitches of the previous row include two double crochets and a chain-1 double-crochet. This is essentially a "filet-like" design: filet crochet is built up of blocks of 4 double-crochets (filled squares in the finished work) and repeats of 2 double crochets separated by 2 chain stitches (open mesh in the finished work) to describe a pattern.

One crochet style that works well with a brick-wall grid is *trellis stitch*, in which arches of chain stitches are anchored to arches of the previous row by single-crochet stitches. Because each arch spans two arches in the previous row, the next row of trellis stitch is offset by half an arch-length. In diagrams and discussions of this process we will, for simplicity, use colors instead of specifying crochet stitches. A block of color can represent actual color or designate a particular stitch or motif. We see the first several rows of trellis designs for Sierpinski triangles and carpets in Figure 7.

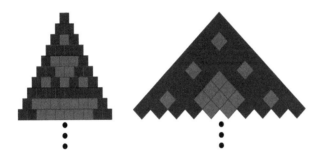

Figure 7. Half-period offset layout for the Sierpinski triangle and carpet.

Note that the Sierpinski carpet design, as constructed here, is actually a design for only half of a Sierpinski carpet; the other half can be constructed by reversing the construction process, if desired. The Sierpinski triangle layout here follows a simplified version of the Campbell design: if a square's previous-row neighbors to the left and right are of the same color, we make the square red; otherwise, we make the square blue. Several interpretations of color as stitches produce pleasing, high-contrast results in this design: the project for

this chapter uses trellis stitch in place of red squares and fan stitches for the blue squares.

Figure 8. The shawl for which instructions are given in Section 4, positioned to match the schema in Figure 7.

2.3 Rule-Based Design and Automata

As we saw in the last section, crochet rules for Sierpinski triangles use a simple method in which the choice of design in a cell depends on the behavior of the cells in the previous row. Such a design methodology has a well-developed foundation in mathematics—it is called a *one-dimensional cellular automaton* (which produces a two-dimensional image if thought of as the progress of a single row of cells over time). Two-state automata are traditionally made up of cells colored black and white, so we will for the time being abandon the blue and red coloring scheme. Also note that automata rules and designs are generally drawn from the top down, unlike crochet diagrams, which are frequently read from the bottom up.

In the elementary one-dimensional cellular automaton, each cell's new state is determined by a rule based on the prior states of the cell and its immediate neighbors [6]. Since each cell can have one of two states, there are $2^3 = 8$ possible state-configurations of three neighboring cells. Each of those state-triples can be associated with one of two new states, so there are $2^8 = 256$

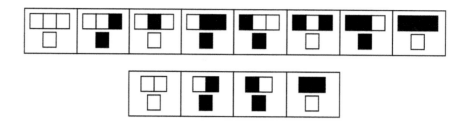

Figure 9. Wolfram's Rule 90 reduces to Mary Pat Campbell's iterative Sierpinski-triangle generation rule.

possible rules. These rules are enumerated by Stephen Wolfram in [6] and [7].

One-dimensional automata are generally well suited for implementation in rectangular-grid crochet designs, and in fact many of the Wolfram rules can generate visually pleasing results when drawn on paper. Rule 90 (shown in Figure 9) is so simple to implement that it has not only been rendered in crochet, but also in self-assembling biological structures [5]; a variant is used with a more complicated start row by Debbie New [4] to create textured knitting, with purl and knit stitches representing the two different cell states. A similarly motif-based approach is suggested by [4], in which single automaton-cells can be represented by 4×4 blocks of cabled stitches.

We can adapt a one-dimensional automata rule to brick-wall-design crochet by omitting the middle square in the rule, because in this type of crochet a stitch is created between two stitches of the previous row. If we represent chain-1 double-crochet pairs of stitches as black squares, and pairs of double-crochet stitches as white squares, then the particular rule for generating the Sierpinski triangle via the Campbell design is shown in Figure 9, using an initial automaton state of a single black square on a white background. The method used to generate the design is equivalent to rule 90 in the Wolfram enumeration, and rules 26, 82, 146, 154, 210, and 218 will in fact also adapt to the same design. While there are 256 one-dimensional automata to be imple-

mented on a rectangular grid, a similar enumeration on the offset grid is disappointing. On this grid, each cell has not three but two parents, so instead of $2^{2^3} = 256$ rules, we only have $2^{2^2} = 16$ rules, most of which do not generate interesting patterns. (Many of the 256 elementary automata are similarly unappealing, but the set of 2-parent rules is too small to allow us to be selective.) There are several possible modifications we could make in order to expand our rule space: we could use more than two colors (which we could render in crochet, for example, by associating three similarly shaped stitches with three different colors), or we could use a wider-ranging network of dependencies, as in having each new cell depend on the four closest cells in the row above, rather than only two. This realm is open to exploration, and there is potential for development of a wide variety of beautiful iterative designs.

3 Teaching Ideas

Many iterative design concepts, such as automata or basic computing theory, lend themselves most easily to enrichment or college classes. On the other hand, Pascal's triangle and binomial coefficients arise explicitly in the secondary curriculum and are involved in the shawl constructed in Section 4. We now expand on these topics in a teaching context.

Crochet provides an excellent visualization project for any module involving discussion of automata. This is partially because crochet itself has one property of automata: each row must be built on the row immediately preceding. A crochet project can thus serve as a challenge to students in crafting automata. Finding algorithms for desired designs is a different sort of challenge.

For instance, while drawing a Sierpinski carpet on graph paper is easy, developing a rule for generating a carpet corner outward as in Figure 7 is surprisingly difficult. To get students started in thinking about how to come up with such a rule, one might start by observing that the first, third, third-to-last and last cells in a row are always the same color, while the second and second-to-last stitches in a row have a color dictated by the row number modulo 3. Coming up with other rules, or a general formula, can challenge students' rule-construction intuition. Alternatively, a simple formula may be derived by coloring a Sierpinski carpet that is not at an angle to the coordinate axes and applying the transformation $(x, y) \mapsto (x + y, x)$ to find a corner-worked layout rule.

In a class that includes computing theory, the question of how to construct a two-dimensional automaton or Turing machine to encapsulate this formula may be a topic for further discussion. Likewise, computational iterative models and recursive formulae can render designs for the dragon curve or the Koch snowflake. In an interdisciplinary curriculum, a module on cellular automata could segue into a discussion of the workings of biological automata (both synthetic and naturally occurring) via the work of Rothemund, Papadakis, and Winfree [5].

Any study of binomial coefficients may be connected to iterative crochet. The generative rule for the offset-grid variant of the Sierpinski triangle is identical to the generative rule for the binomial coefficients mod 2, and the triangle itself is equivalent to Pascal's triangle mod 2. The visual design of the Sierpinski triangle through crochet can help students realize the reasons for this connection: odd terms of the triangle percolate downward along diagonals until they intersect each other and are annihilated. In a pattern such as that used in the shawl below, the movement along diagonals is exhibited more strikingly than inspection of Pascal's triangle would reveal. The visual representation of odd terms cutting through a large expanse of even terms may help students comprehend how Pascal's triangle is related to the generation of Sierpinski triangles. A crochet visualization can lead into a number of open-ended and thought-provoking questions to be asked in the classroom concerning binomial coefficients. For example:

* Why is every term of the $(2^n - 1)$st row of Pascal's triangle odd?

* Does Pascal's triangle exhibit any attractive or predictable patterns considered modulo n for $n \neq 2$? Does it matter whether n is prime? Can we determine viable ways to faithfully represent these patterns in crochet?

* Sierpinski's triangle is revealed by distinguishing between odd and even binomial coefficients; does the design created reveal deeper patterns if we further distinguish between even numbers that are and are not multiples of 4? What about further distinction by multiples of 8? If we factor $\binom{n}{k}$, what is the general formula for the exponent of 2?

For students interested in studying these problems in greater depth, preliminary explorations of the self-similarity expressed in binomial coefficients' factorizations can be found in two papers of Calvin Long [2, 3].

There are also more elementary questions that can be asked of students working with the shawl pattern given in Section 4.

* Which rows will be almost all fans?

* What will the next rows look like? Why?

* How often will the rows with almost all fans occur?

* Which rows will be all arches? Any?

* How many triple-crochets are in row n?

Figure 10. Two views of the Sierpinski shawl.

4 How To Make a Sierpinski Shawl

The shawl in Figure 10 depends on the offset-design method. It uses fans and trellis stitches to depict the Sierpinski triangle on a background of filled and open diamonds.

The Sierpinski shawl is an iterative project, and as such, choices of yarn weight, hook size, and finished project size are actually up to the designer (see Figure 11 for an example), but I recommend working in a fingering or sport-weight yarn with a G or H hook.

Instructions

Row 1: ch 5, dc in first stitch, turn: 1 arch.

Row 2: ch 5, sc in ch-5 arch on row 1, ch 3, dc in same ch-5 arch, turn: 2 arches.

Row 3: ch 5, sc in first arch, ch 5, sc in next ch-5 arch, ch 3, dc in same ch-5 arch: 3 arches.

Row 4: ch 5, sc in first arch, (ch 5, sc in next ch-5 arch) twice, ch 3, dc in same ch-5 arch: 4 arches.

Rows 5–6: ch 5, sc in first arch, (ch 5, sc in each ch-5 arch) to last arch, ch 3, dc in same ch-5 arch. At the end of row 6, there should be 6 arches.

Row 7: ch 5, sc in first arch, (ch 5, sc in next ch-5 arch) twice, 3 tc in next sc (fan made), sc in next ch-5 arch, (ch 5, sc in next ch-5 arch) twice, ch 3, dc in same ch-5 arch: 6 arches and 1 3-tc fan.

Row 8: ch 5, sc in first arch, (ch 5, sc in next ch-5 arch) twice, 3 tc in next sc, sc in middle tc of fan on row 7, 3 tc in next sc, sc in next ch-5 arch, (ch 5, sc in next ch-5 arch) twice, ch 3, dc in same ch-5 arch: 6 arches and 2 3-tc fans.

Row 9: ch 5, sc in first arch, (ch 5, sc in next ch-5 arch) twice, 3 tc in next sc, sc in middle tc of fan, ch 5, sc in middle tc of fan, 3 tc in next sc, sc in next ch-5 arch, (ch 5, sc in next ch-5 arch) twice, ch 3, dc in same ch-5 arch: 7 arches and 2 3-tc fans.

Figure 11. The shawl becomes a kerchief if made with lace-weight yarn.

Row 10: ch 5, sc in first arch, (ch 5, sc in next ch-5 arch) twice, 3 tc in next sc, sc in middle tc of fan, 3 tc in next sc, sc in arch, 3 tc in next sc, sc in middle tc of fan, 3 tc in next sc, sc in next ch-5 arch, (ch 5, sc in next ch-5 arch) twice, ch 3, dc in same ch-5 arch: 6 arches and 4 3-tc fans.

Row 11: ch 5, sc in first arch, (ch 5, sc in next ch-5 arch) twice, 3 tc in next sc, sc in middle tc of fan, (ch 5, sc in middle of tc of next fan) three times, 3 tc in next sc, sc in next ch-5 arch, (ch 5, sc in next ch-5 arch) twice, ch 3, dc in same ch-5 arch: 9 arches and 2 3-tc fans.

Row 12: ch 5, sc in first arch, (ch 5, sc in next ch-5 arch) twice, 3 tc in next sc, sc in middle tc of fan, 3 tc in next sc, sc in next arch, (ch 5, sc in next arch) twice, 3 tc in next sc, sc in middle tc of fan, 3 tc in next sc, sc in next ch-5 arch, (ch 5, sc in next ch-5 arch) twice, ch 3, dc in same ch-5 arch: 8 arches and 4 3-tc fans.

See Figure 12 for a detailed picture of this stitch pattern.

Row 13: ch 5, sc in first arch, (ch 5, sc in next ch-5 arch) twice, 3 tc in next sc, sc in middle tc of fan, ch 5, sc in middle tc of next fan, 3 tc in next sc, sc in next ch-5 arch, ch 5, sc in next ch-5 arch, 3 tc in next sc, sc in middle tc of fan, ch 5, sc in middle tc of fan, 3 tc in next sc, sc in next ch-5 arch, (ch 5, sc in next ch-5 arch) twice, ch 3, dc in same ch-5 arch: 9 arches and 4 3-tc fans.

Row 14: ch 5, sc in first arch, (ch 5, sc in next ch-5 arch) twice, (3 tc in next sc, sc in middle tc of fan, 3 tc in next sc, sc in middle of next arch) four times, (ch 5, sc in next ch-5 arch) twice, ch 3, dc in same ch-5 arch: 6 arches and 8 3-tc fans.

Row 15: ch 5, sc in first arch, (ch 5, sc in next ch-5 arch) twice, 3 tc in next sc, sc in middle of next fan, (ch 5, sc in middle of next fan) 7 times, 3 tc in next sc, sc in next ch-5 arch, (ch 5, sc in next ch-5 arch) twice, ch 3, dc in same ch-5 arch: 13 arches and 2 3-tc fans.

Figure 12. A closeup of the fans and arches.

Rows 16–70: ch 5, sc in first arch. All remaining stitches in the row will be one of two types (fans or arches). To figure out which stitch to do, examine the next two stitches in the previous row—the two stitches your new stitch will touch. If these two stitches are both arches or both fans, then ch 5 and work sc into the middle of the next fan or arch. If one of the two stitches is a fan and the other is an arch, then work 3 tc in next sc and sc in middle of next fan or arch. After the last sc is completed, ch 3 and dc into the same ch-5 arch.

In producing the pattern, be cautious of errors. Be especially careful not to skip stitches in long sequences of ch-5 arches, and examine the overall work frequently. Check that fans only occur adjacent to each other in even-numbered rows, and that rows 14, 22, 38, and 70 have long rows of fans flanked on each end by 3 arches.

Row 71: ch 5, sc in first arch, (ch 5, sc in next ch-5 arch) twice, (3 tc in next sc, sc in middle tc of fan) 64 times, 3 tc in next sc, sc in next ch-5 arch, (ch 5, sc in next ch-5 arch) twice, ch 3, dc in same ch-5 arch: 6 arches and 65 3-tc fans.

Row 72: ch 5, sc in first arch, (ch 5, sc in next ch-5 arch) twice, (ch 5, sc in middle tc of fan) 65 times, (ch 5, sc in next ch-5 arch) three times, ch 3, dc in same ch-5 arch: 66 arches.

Bibliography

[1] Campbell, Mary Pat. "Fractal Crochet." http://www.marypat.org/stuff/nylife/020325.html, March 25, 2002. Also "Pattern for Fractal Shawl," http://community.livejournal.com/crochet/9822.html, March 25, 2002.

[2] Long, Calvin T. "Pascal's Triangle modulo *p*." *Fibonacci Quarterly*, vol. 19, no. 5, 1981, pp. 458–463.

[3] Long, Calvin T. "Some Divisibility Properties of Pascal's Triangle." *Fibonacci Quarterly*, vol. 19, no. 3, 1981, pp. 257–263.

[4] New, Debbie. "Cellular Automaton Knitting." *Knitter's Magazine*, vol. 49, 1997, pp. 82–83.

[5] Rothemund, Paul W. K., Papadakis, Nick, and Winfree, Erik. "Algorithmic Self-Assembly of DNA Sierpinski Triangles." *PLoS Biology*, vol. 2, 2004, e424.

[6] Wolfram, Stephen. *A New Kind of Science*. Wolfram Media, Champaign, IL, 2002.

[7] Wolfram Institute. *The Wolfram Atlas of Simple Programs*. http://atlas.wolfram.com, 2007.

CHAPTER 4

only two knit stitches can create a torus

SARAH-MARIE BELCASTRO

1 Overview

Not everyone knits the same way. There is a standard right-handed knit stitch, which is standard because every knitting instruction book in the Western world uses it. Still, in any random collection of knitters there will be a few who knit differently. There are a couple of variations commonly encountered on left-handed knit stitches, and some knitters use what's technically known as a twist stitch as their basic stitch. It would be easy to conclude that most knitters use the same stitch because it's the one they learned from books or relatives, and that some people interpreted book instructions differently or had a different stitch handed down in their families. There's still an unanswered question lurking about, though. Why did the standard knit stitch become standard? That is, was there a reason, or did it just happen to be passed on?

My knitting history provides a clue to the answer, although it turns out to be a somewhat misleading clue. I learned to knit while in elementary school. My mother is right-handed and I am left-handed, and so the simplest way for her to teach me to knit was to face me. (Most people sit side by side when teaching and learning knitting or crocheting, because *most* people are right-handed.) The methodology I picked up in this fashion served me well for many years—in high school, I knit my way through a 14-foot scarf, and in the process taught myself to knit ambidextrously. I ceased knitting until my senior year of college, but then proceeded to design and knit several sweaters before noticing, several years into graduate school, that there was something . . . wrong. When I attempted complicated stitches (those involving more than just knitting and purling) from traditional knitting books, the stitches did not come out correctly. They were either more or less twisted than pictured in books. After first thinking that the error was in my interpretation of the instructions, I finally realized that something more significant was happening: something was fundamentally different about my left-handed way of knitting from the conventional right-handed way of knitting. Somehow my knitting methodology was fine for simple stitches, but not for complex ones. After a while, I sat down and tried to learn to knit right-handedly from the beginning—that is, by using the instructions for the standard knit stitch. This was quite an eye-opening experience and led to the research described here.

In Section 2, I use very basic combinatorics (counting), observations of symmetry, and brute force to address the following questions:

* How many different ways are there to construct a knit stitch? Are any of these ways equivalent to each other in terms of the resulting fabric?

* How many different ways are there to construct plain-knitted (stockinette) fabric? How many different plain-knitted fabrics are there?

* Knit and purl stitches are not symmetric in the way we execute them, but the resulting fabric is symmetric. Does there exist a symmetric execution of knitting and purling?

and, finally,

* Of the different knit stitches, why did Western culture settle on the stitch now considered traditional?

For the mathematicians reading this, Section 2 also contains both mathematical and practical descriptions of knitting.

Here is the quick-and-informal version of the answers to the above questions:

* There are eight possible ways to make a knit stitch, which correspond to all possible combinations of places to insert the needle into the loop, directions to point the needle through the loop, and directions to wrap the yarn around the needle. Four of these possibilities turn out to be increases rather than knit stitches, so we discard them. That leaves us with four different ways to make a knit stitch.

* We can construct fabric with each of the four knit stitches by knitting in the round. If we take the four different knit stitches and look at them from the other side of the work, they become four purl stitches. Using regular flat knitting, we have to have a knit stitch and a purl stitch to make stockinette fabric, so we check all sixteen possible combinations of those four knit and four purl stitches. There are six different stockinette fabrics that result from those twenty possible fabric constructions.

* There are two stitch pairs among the sixteen possible, each with one knit stitch and one purl stitch, that are symmetric both in execution and in result.

* Of the four knit stitches, exactly two work the same way for flat knitting as they do for circular knitting; one of these is the standard knit stitch. This property is likely the reason why it has become the standard stitch. The other stitch that functions consistently in the round and for flat knitting is (in essence) a reflection of the standard knit stitch. Interestingly, I use the reflected knit stitch—this is why my attempts at complex stitches induce odd twistings of the yarn strands. I would need to apply the same sort of reflection to the instructions in order to get the desired (and expected) results, and I had always interpreted the instructions using a different reflection.

That last discussion leaves something out: *why* would one need to use both circular and flat knitting in one project? It's efficient to avoid seams when possible, as in a sweater knitted from the bottom up. Some sweater designs require circular knitting up to the armpits, and then split into a front section of flat knitting and a back section of flat knitting which come together again at the shoulders. Similarly, one knits socks in the round but uses flat knitting in order to construct the heels. There is also a mathematical object that must be knitted using both circular and flat knitting, and that is the project presented in this chapter: a torus, or in lay terms, the surface of a doughnut. The simplest way to knit a torus is to create a curved cylinder and join the ends. Circular knitting is used to make the cylinder; the reason flat knitting is also needed is that we use "short rows" in order to give the cylinder curvature. Short rows are the technique used to turn the heel of a sock, and they basically insert more stitches on one side of a cylinder than the other; in order to do this, the circular knitting is paused while flat knitting adds rows with fewer stitches (thus short rows).

There are more details on all of the above topics in the following sections—so read on!

2 Mathematics

Before delving into a mathematical analysis of the knit stitch, I am going to give a mathematical description of the process of knitting. While knitters will intuitively know everything I say in the following section, they may want to read it in order to understand the terminology I use in the subsequent mathematical analysis.

2.1 About Knitting

Knitting is a method of constructing fabric from yarn. No knots are introduced in the process of knitting, so aside from anchoring knots made at the beginning and end of a piece, any knitted object is equivalent to the unknot. Figure 1 shows the path yarn takes through a small section of knitting. (The knitting-savvy reader may notice

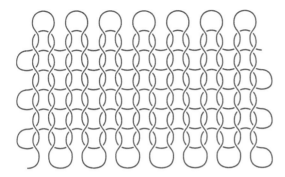

Figure 1. A small section of left-handed flat knitting.

that this is a left-handed piece of knitting, as the free ends of the yarn are at top right and bottom left.)

A knitter creates one stitch at a time, which appears as a new loop of yarn interlocking both horizontally and vertically adjacent loops that were created earlier. She holds one needle in each hand; these may be two separate, straight needles or two needles joined by a cable.

One of the needles (the left needle if one is right-handed, or the right needle if one is left-handed) holds loops that have already been made. We'll call that the *passive* needle.

To create a new loop, a knitter pokes the other needle (the *active* needle) into the nearest loop on the passive needle, wraps the free end of the yarn once around it (which makes an unattached loop), and pulls the unattached loop through the nearest loop on the passive needle (usually slipping it off the passive needle). This creates a new loop on the active needle. Continuing to do this creates a new row of loops and transfers the knitting onto the active needle. When all loops have been transferred from the passive needle to the active needle, a row is complete. In order to continue, the work must be turned around, reversing the roles of the needles. (In the case of two needles connected by a cable, loops are created on the active needle and passed along the cable to the passive needle for continuous knitting pleasure.) In each of the figures in Section 2.2, the left-hand diagram has the active needle on the left, and the right-hand diagram has the active needle on the right.

That is, left-handed people will prefer the diagrams in the left halves of the figures, and right-handed people will prefer the diagrams in the right halves of the figures.

2.2 Combinatorial Analysis

Let's look at the different ways there are to construct a knit stitch. This necessitates a definition.

Definition 1 A *knit stitch* is any stitch that creates exactly one new loop on the active needle from one loop on the passive needle.

We consider two knit stitches to be different if the knitter uses a different set of motions with the yarn and/or needles for each stitch. Now, let us enumerate the various motions that are possible in forming a knit stitch.

⋆ The free end of the yarn may either be held in back of the work, denoted y_b (for *yarn-back*, see Figure 2), or in front of the work, denoted y_f (for *yarn-front*, see Figure 3). Note that one could also hold the free end of the yarn in either the active-needle hand or the passive-needle hand. This changes only the angle of the yarn to the work, and does not affect the structure of the stitch. The style of holding the yarn in the passive-needle hand is called Continental knitting whereas that of holding the yarn in the active-needle hand is called English knitting.

⋆ The active needle may be inserted into the front of the loop, denoted i_f (for *insertion-front*, see Figure 3), or into the back of the loop, denoted i_b (for *insertion-back*, see Figure 2). It seems at first as though each choice will produce the same result, because after inserting the active needle on one side of the loop, we might maneuver the point of the active needle around the passive needle while keeping it through the loop. While i_f appears topologically equivalent to i_b, this is not so. The resulting stitches differ because

Left-Handed View   **Right-Handed View**

Figure 2. A needle inserted into a loop, y_b, i_b, i_p.

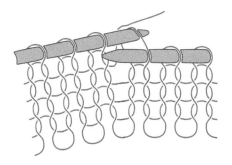 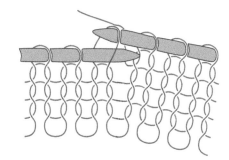

Figure 3. A needle inserted into a loop, y_f, i_f, i_ℓ.

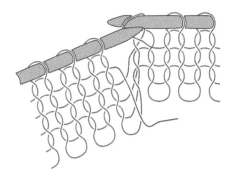 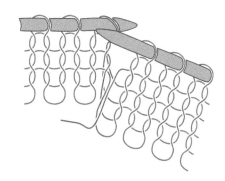

Figure 4. The top diagrams show *under* and the bottom two diagrams show *over*.

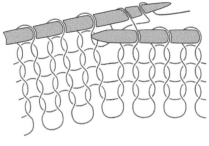 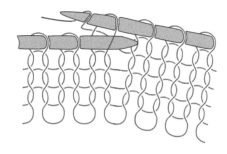

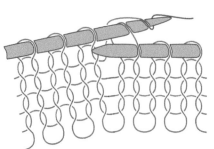 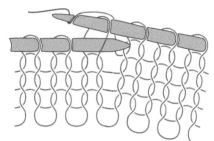

the free end of the yarn needs to cross the active needle in one case but not the other. (Which case is which depends on whether the free end of the yarn is y_b or y_f.)

★ Similarly, the active needle may be inserted from the *needle-point side* of the loop, denoted i_p (see Figure 2), or from the *needle-length side* of the loop, denoted i_ℓ (see Figure 3).

★ The yarn is wound once around the active needle, either under the needle, or over the needle. By the term *under*, we mean that the yarn is pulled under the needle to the other side of the work, then over to the original side of the work; the term *over* means the yarn is pulled up over the needle to the other side of the work, then wound under to the original side of the work. See Figure 4 for examples.

There are four aspects to a knit stitch; with two choices for each, we obtain $2^4 = 16$ combinations. But we can use symmetry to bring that number down to eight—each y_f stitch is a y_b stitch viewed from the other side of the work. This symmetry operation, that of turning the work around, pairs the variables as follows:

y_b	i_p	i_ℓ	i_b	i_f	over	under
y_f	i_p	i_ℓ	i_f	i_b	over	under

At this point, the reader may wish to pick up yarn and knitting needles in order to verify the following observations. If we attempt each of the remaining eight combinations (this is known as proof by exhaustion), we discover that four of them are increases, rather than knit stitches. That is, they create two loops on the active needle rather than one. Those four are the two i_p/i_f combinations and the two i_ℓ/i_b combinations. In these increases, the extra loop is created because the yarn has to cross the active needle behind the loop on the passive needle before wrapping around the active needle.

That leaves us with four acceptable knit stitches:

$$\begin{array}{lll} \text{Let} & [k_1] & \text{be} \quad y_b/i_p/i_b/\text{over}, \\ & [k_2] & \phantom{\text{be}} \quad y_b/i_p/i_b/\text{under}, \\ & [k_3] & \phantom{\text{be}} \quad y_b/i_\ell/i_f/\text{over}, \\ & [k_4] & \phantom{\text{be}} \quad y_b/i_\ell/i_f/\text{under}. \end{array}$$

The traditional knit stitch is $[k_4]$, and the author uses $[k_1]$.

Now we need to decide whether any of these four stitches are equivalent in terms of the resulting fabric. There are two ways to create a plain-knitted fabric:

1. Use circular needles, knit, and create a cylinder by going around and around. There are four ways of doing this, as there are four acceptable knit stitches.

2. Knit one row, turn the knitting, and purl the following row; thereafter alternate knit rows and purl rows.

Definition 2 A *purl stitch* is a y_f stitch that corresponds to one of the knit stitches (y_b stitches) under the symmetry operation of turning the work. We denote by $[p_i]$ the stitch $[k_i]$ viewed from the other side of the work.

For circular knitting, $[k_1]$ and $[k_4]$ each gives the standard knit fabric. $[k_2]$ results in twisted stitches. $[k_3]$ results in stitches twisted the opposite way from $[k_2]$. In the literature, unsurprisingly, $[k_2]$ is called "Knit stitch—crossed" and $[k_3]$ is called "Purl stitch—crossed" [3]. See Figure 5 in Section 2.4.

This suggests that $[k_1]$ and $[k_4]$ might be functionally equivalent. To determine whether this is the case, we need to check the results of using each stitch for flat knitting along with its paired purl stitch. It turns out (see Section 2.4) that they each produce twisted-stitch fabric, with alternating rows twisted in opposite directions. The two fabrics are identical up to a vertical shift of one row, so we may consider them equivalent.

To explain why this is the case, as well as to provide background for our subsequent classification of all plain-knitted fabrics, we will now analyze the structure of knitted stitches.

2.3 Structural Analysis of Stitch Interactions

There are two ways a completed stitch may hang on a needle. The half of the stitch closest to the needle point may drape over the front or over the back of the needle. If we use the terminology of [2], we say that stitches hang *nearside* or *farside* respectively. Which way a stitch hangs depends on the way in which the yarn is wound around the needle; wrapping the yarn over results in stitches hanging nearside, and wrapping the yarn under results in stitches hanging farside. Note that in both circular and flat knitting, nearside stitches on the active needle appear farside when they are on the passive needle (see the figures in Section 2.2).

Whether completed knit stitches are twisted, and, if so, which way, depends on the way the stitches hang on the passive needle and on the construction of the stitches in the subsequent row. That is, if one knits into the back of a nearside stitch (i_p/i_b), the side of the stitch closest to the point of the needle will twist in front; if one knits into the front of a farside stitch (i_ℓ/i_f), the side of the stitch further from the point of the needle will twist in front. Again, the reader may find that experimenting with yarn and needles will make this clearer. A full explanation of how twisting functions relative to the standard knit stitch appears in [2, pp. 23–25 and 32–35]. In order to avoid handed references, let us describe the side of a stitch (on the passive needle) closest to the point of the needle as "active" and describe the other side of the stitch as "passive." Thus the twisting of a stitch will be abbreviated as a/p (for active side twisted over passive side) or p/a (for passive side twisted over active side).

Given this analysis, the circular-knitting results make sense. For example, $[k_1]$ produces nearside stitches on the active needle which become farside stitches on the passive needle. Knitting subsequent rows with $[k_1]$ means we knit into the back of each (now farside) stitch, so the resulting stitches are not twisted. On the other hand, $[k_2]$ produces farside stitches on the active needle which become nearside stitches on the passive needle; knitting subsequent rows with $[k_2]$ means we knit into the back of each (now nearside) stitch, so the active sides of the stitches will twist over the passive sides of the stitches in the finished fabric.

2.4 Results

There are four knit stitches and four purl stitches, and thus 16 knit-purl stitch combinations. Twelve of these combinations appear to form six equivalent pairs under the symmetry operation of turning the work. (For example, from the definitions we would expect the pair $[k_1] - [p_4]$ to be equivalent to $[p_1] - [k_4] = [k_4] - [p_1]$.) However, the symmetry operation also reverses stitch-twisting, so the pairs will only be equivalent when no twisting is present.

We predict the following results for flat knitting, which the reader can verify experimentally:

$[k_1]$ alternated with $[p_1] \rightarrow$ twisted a/p alternating with twisted p/a

$[k_1]$ alternated with $[p_2] \rightarrow$ twisted a/p alternating with no twist

$[k_1]$ alternated with $[p_3] \rightarrow$ no twist alternating with twisted p/a

$[k_1]$ alternated with $[p_4] \rightarrow$ standard knit (all rows)

$[k_2]$ alternated with $[p_1] \rightarrow$ no twist alternating with twisted p/a

$[k_2]$ alternated with $[p_2] \rightarrow$ standard knit (all rows)

$[k_2]$ alternated with $[p_3] \rightarrow$ twisted p/a (all rows)

$[k_2]$ alternated with $[p_4] \rightarrow$ twisted p/a alternating with no twist

$[k_3]$ alternated with $[p_1] \rightarrow$ twisted a/p alternating with no twist

$[k_3]$ alternated with $[p_2] \rightarrow$ twisted a/p (all rows)

$[k_3]$ alternated with $[p_3] \rightarrow$ standard knit (all rows)

$[k_3]$ alternated with $[p_4] \rightarrow$ no twist alternating with twisted a/p

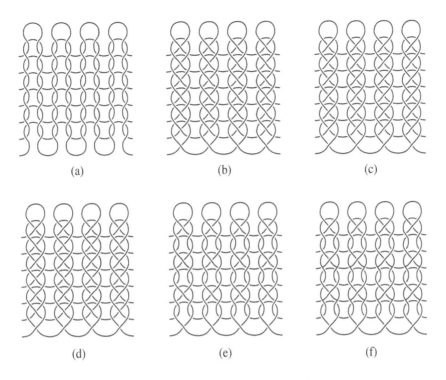

Figure 5. The six possible knitted fabrics as listed in Theorem 3.

$[k_4]$ alternated with $[p_1] \rightarrow$ standard knit (all rows)

$[k_4]$ alternated with $[p_2] \rightarrow$ no twist alternating with twisted a/p

$[k_4]$ alternated with $[p_3] \rightarrow$ twisted p/a alternating with no twist

$[k_4]$ alternated with $[p_4] \rightarrow$ twisted p/a alternating with twisted a/p

Another way of looking at this data is Theorem 3, whose proof follows from Section 2.3 and which was verified above.

Theorem 3 *There are (up to a single vertical row shift) six different fabrics as pictured in Figure 5, four of which can be created in multiple ways:*

(a) *Standard knit can be created by the pairs $[k_1] - [p_4]$, $[k_2] - [p_2]$, $[k_3] - [p_3]$, and $[k_4] - [p_1]$.*

(b) *All rows twisted p/a can only be created by the pair $[k_2] - [p_3]$.*

(c) *All rows twisted a/p can only be created by the pair $[k_3] - [p_2]$.*

(d) *Rows twisted alternating between p/a and a/p can be created by the pairs $[k_1] - [p_1]$ and $[k_4] - [p_4]$.*

(e) *Alternate rows twisted a/p and untwisted can be created by the pairs $[k_3] - [p_1]$, $[k_1] - [p_2]$, $[k_3] - [p_4]$, and $[k_4] - [p_2]$.*

(f) *Alternate rows twisted p/a and untwisted can be created by the pairs $[k_2] - [p_4]$, $[k_4] - [p_3]$, $[k_2] - [p_1]$, and $[k_1] - [p_3]$.*

From the above analysis, we can answer another of our original questions. The traditional knit and purl stitches are not executed symmetrically. Does there exist a symmetric execution of knitting and purling that produces standard knitted fabric? The symmetric pairs $[k_2] - [p_2]$ and $[k_3] - [p_3]$ both produce standard knitted fabric. Personally, I think $[k_2] - [p_2]$ is easier to do and remember.

We now address the question that motivated this work. Mary Thomas [3, pp. 50–57] notes that the traditional knit stitch has been the same in Western Europe since the time of the Middle Ages. Of the different knit stitches, why did Western culture settle on the stitch now considered traditional? The big punch line is that we knit the way we do for a reason; it's one of two possible ways to knit that works for both circular and double-needle knitting, in the sense that it produces standard knitted fabric. Each of $[k_1]$, $[k_4]$ (likewise $[p_1]$, $[p_4]$) can be used in both flat and circular knitting without twisting. (Aside: I have actually met people who do $[k_2]$ or $[k_3]$. They either choose never to knit in the round, or they fix the stitches as they go, by un-twisting them just before they knit into them when they do knit in the round.) As for why Western culture uses stitch $[k_4]$ rather than $[k_1]$, I can only conjecture that the "under" motion may be more physically efficient than the "over" motion.

It is interesting to note the relationship between $[k_1]$, $[k_4]$ and handedness. Traditionally, left-handers do $[k_4]$, by mirroring left–right the motions of a right-hander. (This is how it's taught in books.) In fact, $[k_1] - [p_4]$ is really $[p_1] - [k_4]$ mirrored front–back, or in other words, what you learn to do if you're facing someone knitting traditionally and switch the knit and purl.

2.5 Understanding the Torus

Seasoned mathematicians may wish to skip this section (or, conversely, perhaps be comforted by reading a section they understand completely). Technically, a torus is a topological surface, a smooth compact two-dimensional manifold without boundary. This means that when we look at an instantiation of the torus in regular three-dimensional space (see Figure 6 (left)), we see that the torus has no corners or pinch-points, that the torus is finite in size, that the torus is exactly the bound-

ary of the shape it encloses, that we may impose coordinates (equivalently, draw perpendicular axes) on any small patch of the torus, that we can translate such coordinates across overlapping patches, and that it has no edges.

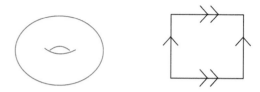

Figure 6. A torus in 3-space (left). A torus drawn in the plane (right).

Unfortunately, the usual representation of a torus does not show all sides of the surface. Thus, we have an alternate representation, shown in the right-hand diagram in Figure 6; it is made by slicing through the torus to create a cylinder and slicing the cylinder lengthwise to create a rectangle. This process is shown in Figure 7.

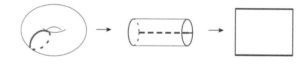

Figure 7. Transforming a torus in three dimensions to one in two dimensions.

3 Teaching Ideas

Possible uses of this material range from the elementary school classroom to the university. The mathematical ideas used in the knitting analysis are elementary combinatorics, identification of symmetry, brute force/exhaustion, and breaking a problem down into parts.

In a Waldorf or Montessori school, where knitting may be part of the curriculum, it would be natural to have students experiment with different constructions for knit and purl stitches as a way of discovering why

Figure 8. Three tori.

the traditional knit stitch is constructed as it is. Likewise, students might be asked to look for various symmetries in knitted fabric; advanced students could look at the knit/purl transformation and contrast it to the left/right transformation.

Mathematics for Liberal Arts courses could use the eight possible knit-stitch constructions (see Section 2.2) as a very elementary example of combinatorics, as well as an example of the use of exhaustion as an analysis technique.

A course in Mathematical Modelling might look at this material as an example of how to approach practical problems in a mathematical way. The mathematical description of a knit stitch gives an example of breaking a problem into smaller, manageable parts. The attempt to make all eight stitches indicated by the combinatorics shows that not everything that appears as a mathematical solution is also a physical one (because the increases that result are not knit stitches). Observing the knit/purl work-reversing symmetry gives an example of how using symmetries can simplify the analysis of a problem.

This chapter's project can also be used in the classroom. One obvious role is as a physical example of a torus, for any level of mathematics class.

There are several questions about the addition of texture which might be asked of high-school students or Mathematics for Liberal Arts students. First, we pose two questions that are relatively easy to answer once students have understood the pattern for a generalized plain torus (given in Section 4.2).

Question 1: Given a texture-pattern we'd like to knit onto a torus (such as those given in Figure 9), how should we alter the number of stitches per round?

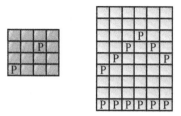

Figure 9. Two of the textures knitted onto tori pictured in this chapter. The capital "P"s indicate purl stitches.

Answer: We should change the number of stitches per round to be an integer multiple of the pattern repeat, e.g. a multiple of 4 or a multiple of 6 for the two patterns shown in Figure 9.

Question 2: How many times will we repeat a texture pattern of *k* rows before finishing the torus?

Answer: If the circumference of the curved cylinder that makes the torus is *c* inches, and the gauge is *x* stitches by *y* rows, then there will be $\lceil c \cdot y/k \rceil$ repeats of the texture pattern. (This unusual-looking notation is for the ceiling function, which returns the least integer greater than the input, or rounds a fractional input up to the next integer.)

There are also some questions that are significantly more difficult to answer.

Question 3: A texture pattern is designed on a flat, uncurved section of knitting. How can we extend this pattern to the parts of the torus that involve the slipped/wrap-and-turned stitches?

Answer: Consider an extended version of the texture-pattern chart that takes into account these stitches, as in Figure 10.

A part of the solution for the zig-zag pattern takes into account the interaction between the slipped/wrap-and-turn stitches and the starting column of the chart. Note that the pattern has been shifted two columns to the right in order to mesh the slipped/wrap-and-turn stitches into the pattern. (Trial and error shows that there exist an equivalent but mirrored shift and two less-optimal shifts.)

Question 4: Can we extend the texture patterns onto the interior of the torus in such a way as to make the transition appear seamless? If so, how?

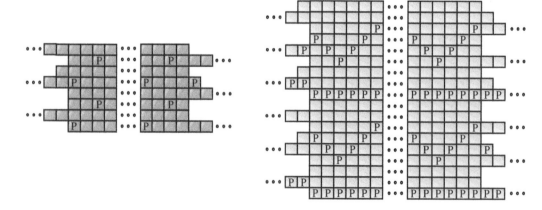

Figure 10. Altering a pattern near the "rectangle" boundaries.

Answer: Whether or not this is possible depends highly on the texture pattern. For example, the zig-zag pattern has nine rows; because each row of the torus interior corresponds to three rows on the torus exterior, each exterior copy of the zig-zag pattern produces three interior rows—not enough space to do more than extend the horizontal border into one of the rows. There would need to be a zig-zag pattern of height 12 in order to have a single horizontal stripe extending the zig-zag. On the other hand, the pattern of isolated bumps can be extended, but choosing where to place the bumps is a challenge. The pattern is four rows high, which is relatively prime to (has no common factors other than 1 with) the exterior/interior row ratio of 3, so 12 exterior rows will be knitted before any added interior pattern repeats.

The torus is also useful in topology/geometry classes, as a way of thinking about how to approximate embedding curvature using a discrete medium. Students might be asked to create other ways to mimic curvature discretely, and to compare the associated physical and visual results and the varying complexities in construction.

Question 5: How could the plain torus pattern be altered to change the ratio of the longitudinal to meridianal cycles?

Figure 11. A differently-proportioned doughnut.

Answer: Change the short-row pattern. For example, in the torus in Figure 11 there are two short rows created for each complete knitted round. In the plain torus for which instructions are given in Section 4, a single short row is created for each complete knitted round.

4 How to Make a Torus

There are *so* many design possibilities in making a torus. The instructions here are for plain tori, but one could add color or texture as well. As you will see in making the plain torus, it is created using short rows to form an outer "rectangle" linked by wrap-and-turn stitches to a shorter inner "rectangle." Almost any texture or color scheme can easily be transferred to the "rectangles," but for a successful torus one needs to pay attention to the way the patterns of color/texture blend along the boundaries between the "rectangles." See [1] for an example of color and Figures 13 and 8 for examples of texture.

4.1 One Particular Torus

Materials

* Reynolds Saucy, one skein; a few yards of scrap yarn in a contrasting color.

* A 9″ size 7 circular needle (Clover makes these) or a set of size 7 double-pointed needles.

* Two stitch markers.

* Stuffing (such as polyester fiberfill) if you want a 3D torus as the result!

Gauge

* 30 stitches / 7 ″ and 6 rows/1 ″.

Figure 12. Tori make lovely snacks.

Instructions

The only special procedure needed is a wrap and turn, used to avoid holes appearing when doing short rows (inducing curvature).

Knit row wrap and turn: slip next stitch knitwise, bring yarn to front, turn work, slip same stitch purlwise, bring yarn to front.

Purl row wrap and turn: slip next stitch purlwise, bring yarn to back, turn work, slip same stitch knitwise, bring yarn to back.

Getting Started
Using chain cast-on, cast on 30 st with scrap yarn.
K 30 with primary yarn.
Join stitches into a circle and K 20.
Add stitch marker. Wrap and turn. (If using double-pointed needles, move some stitches to adjacent needles as necessary so that the stitch markers don't fall off.)
P 20.
Add stitch marker. Wrap and turn. (If using double-pointed needles, be sure to move stitches to the needle so the stitch marker doesn't fall off.)

K 20, pick up wrapped strand with next stitch and knit together.
K 8, pick up wrapped strand with next stitch and knit together.

The Basic Repeat
(1) K 19, wrap and turn on stitch before marker.
(2) P 18, wrap and turn on stitch before marker.
(3) K 18, pick up wrapped strand with next stitch and knit together.
(4) K 10, pick up wrapped strand with next stitch and knit together.
(5) K 19, wrap and turn.
(6) P 20, wrap and turn.
(7) K 20, pick up wrapped strand with next stitch and knit together.
(8) K 8, pick up wrapped strand with next stitch and knit together.

Most of the Torus
Repeat (1)–(8) 19 times total, so that there are 40 rows on the shorter side of the cylinder.

If you knit tightly, then your torus may not be close to complete at this point. Lay it flat to see whether the shorter side nearly makes a circle and the longer side makes $\approx$ 6 rows short of a circle. If 1/6 or more of the circumference is missing, repeat (1)–(8) again.

Finishing Up

K 19, wrap and turn on stitch before marker.

P 18, wrap and turn on stitch before marker.

K 18, pick up wrapped strand with next stitch and knit together.

K 10, pick up wrapped strand with next stitch and knit together.

K 20, wrap and turn.

P 20, wrap and turn.

K 20.

Stuff the torus to desired fullness. Graft using kitchener stitch, being sure to pick up the two wrapped strands when grafting their corresponding stitches. Because the cast-on was chained, you should be able to release the stitches as you graft, so that the scrap yarn can be easily removed as the torus closes up.

Weave in ends.

4.2 Generalized Torus Construction

In Section 4.1 instructions are given for a 30-stitch circumference torus. Here are instructions for a torus of any circumference.

Materials

* A skein of yarn; a few yards of scrap yarn in a contrasting color.

* A circular needle, approximately the length of the circumference (c inches) of a circular cross-section of the desired torus, or a set of double-pointed needles.

* Two stitch markers.

* Stuffing (such as polyester fiberfill) to 3D-ify the torus.

Determine the gauge of the yarn on the selected needles, and denote this x stitches per inch and y rows per inch.

Instructions

The only special procedure needed is a wrap and turn, used to avoid holes appearing when doing short rows (inducing curvature).

Knit row wrap and turn: slip next stitch knitwise, bring yarn to front, turn work, slip same stitch purlwise, bring yarn to front.

Purl row wrap and turn: slip next stitch purlwise, bring yarn to back, turn work, slip same stitch knitwise, bring yarn to back.

Getting Started

Using chain cast-on, cast on $x \cdot c$ st with scrap yarn. Denote this number of stitches by $x \cdot c = r$.

K r with primary yarn. If using double-pointed needles, divide the stitches unevenly between the needles so that neither stitch marker will be on an end of a needle. Join stitches into a circle and K $2r/3$. If r is not evenly divisible by 3, round $r/3$ to the nearest integer k and K $r - k$ stitches instead.

Add stitch marker. Wrap and turn.

P $2r/3$.

Add stitch marker. Wrap and turn.

K $2r/3$, pick up wrapped strand with next stitch and knit together.

K $r/3 - 2$, pick up wrapped strand with next stitch and knit together.

The Basic Repeat

(1) K $2r/3 - 1$, wrap and turn on stitch before marker.

(2) P $2r/3 - 2$, wrap and turn on stitch before marker.

(3) K $2r/3 - 2$, pick up wrapped strand with next stitch and knit together.

(4) K $r/3$, pick up wrapped strand with next stitch and knit together.

(5) K $2r/3 - 1$, wrap and turn.

Figure 13. Even cats can be lovers and collectors of tori.

(6) P $2r/3$, wrap and turn.

(7) K $2r/3$, pick up wrapped strand with next stitch and knit together.

(8) K $r/3 - 2$, pick up wrapped strand with next stitch and knit together.

Most of the Torus

Repeat (1)–(8) $(c \cdot y - 4)/2$ times total, so that there are $c \cdot y - 2$ rows on the shorter side of the cylinder.

If you knit tightly, then your torus may not be close to complete at this point. Lay it flat to see whether the shorter side nearly makes a circle and the longer side makes ≈ 6 rows short of a circle. If 1/6 or more of the circumference is missing, repeat (1)–(8) again.

Finishing Up

K $2r/3 - 1$, wrap and turn on stitch before marker.

P $2r/3 - 2$, wrap and turn on stitch before marker.

K $2r/3 - 2$, pick up wrapped strand with next stitch and knit together.

K $r/3$, pick up wrapped strand with next stitch and knit together.

K $2r/3$, wrap and turn.

P $2r/3$, wrap and turn.

K $2r/3$.

Stuff torus to desired fullness. Graft using kitchener stitch, being sure to pick up the two wrapped strands when grafting their corresponding stitches. Because the cast-on was chained, you should be able to release the stitches as you graft, so that the scrap yarn can be easily removed as the torus closes up.

Weave in ends.

Bibliography

[1] belcastro, sarah-marie and Yackel, Carolyn. "The Seven-Colored Torus: Mathematically Interesting and Nontrivial to Construct." In *Homage to a Pied Puzzler*, edited by Ed Pegg Jr, Alan H. Schoen, and Tom Rodgers, to appear. A K Peters, Wellesley, 2007.

[2] Hiatt, June Hemons. *The Principles of Knitting*. Simon and Schuster, New York, 1988.

[3] Thomas, Mary. *Mary Thomas's Knitting Book*. Dover, New York, 1938; reprinted 1972.

CHAPTER 5

symmetry patterns in cross-stitch

MARY D. SHEPHERD

1 Overview

Symmetry occurs all around us in faces and flowers and snowflakes. The craftwork and art of most cultures involves symmetry, often in the design of backgrounds, borders, and focal points. Symmetry involves the repetition of a basic pattern. The most common repetition is called a *translation*, in which the basic pattern moves along a line by a specific amount, leaving the pattern unchanged. We can use translations to classify symmetry patterns into three broad categories. A *rosette* pattern has at least one point that is not moved by any of the symmetry transformations, and thus has no translations. *Frieze* patterns, frequently used for fabric and paper borders, have translations in two directions along one line. *Wallpaper* patterns have translations in each direction along two intersecting lines.

In counted cross-stitch, one works on evenweave fabric (so called because it has the same number of threads per inch vertically and horizontally). The basic stitch appears as an × on the fabric (instructions on how to cross-stitch are given in Section 4 of this chapter). There are two common types of evenweave fabric used in counted cross-stitch: linen and Aida. In linen cloth each thread is counted individually, and a single cross-stitch is done over two threads (see Figure 1).

On Aida cloth (see Figure 2), threads are counted in groups, and a cross-stitch is done over one thread grouping. Aida fabric is easier to use, particularly for beginners, since one counts the blocks of threads and not individual threads.

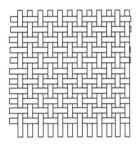  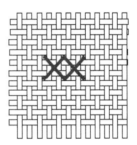

Figure 1. Example of evenweave linen fabric, schematically on the left, pictorially in the middle, and with two cross-stitches shown on the right.

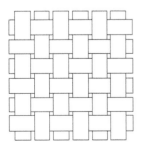  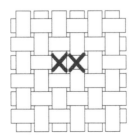

Figure 2. Example of Aida fabric, schematically on the left, pictorially in the middle, and with two cross-stitches shown on the right.

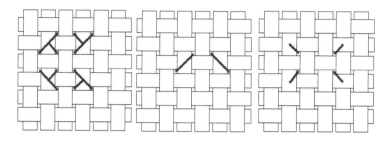

Figure 3. Demonstration of each of the possible 3/4, half, and 1/4 stitches.

Stitching over squares constrains the number of symmetry patterns that are possible. Why is this? If one thinks of the fabric as a grid of squares like graph paper, the amount covered by a stitch is a single square. The only possible subdivisions of this square are with stitches that "cover" half a square on the diagonal (see Figure 3), namely the 3/4 stitch (which uses three of the four corners of the square), the half stitch (which uses two diagonal corners of a square), and the 1/4 stitch (which uses one corner of the square and the center of the square). The 1/4 stitch is often combined with an outline stitch on the diagonal so that it appears as a 3/4 stitch when the outline stitch is included.

All of these will be referred to as half cross-stitches in this chapter, as each covers half of a square bounded by a diagonal of the square. Thus our covering stitches are all composed of either unit squares or unit-sided isosceles right triangles. This means that the only angles one can create with the covering stitches in a counted cross-stitch pattern are multiples of 45°, the smaller angle in the half cross-stitch. See Figure 4 for examples. Even outline stitches are constrained because the thread must go from a corner of a square to a corner of the same or another square, so the slope (the ratio of the number of squares up or down to the number of squares left or right) of these outline stitches is always a fraction.

In mathematical terms, the tangent of the angle made by outline stitches must be a rational number. This means that two of the angles common in wallpaper patterns, 60° and 120°, are not possible. It is possible to get a fairly close approximation of 60° with outline stitches, but we are only interested in exact angles.

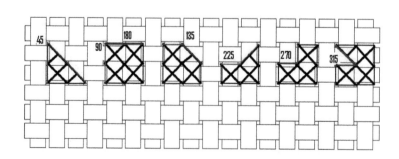

Figure 4. Angles possible with covering stitches.

1.1 The Basic Transformations

Before we look at the different types of patterns possible, let's discuss the four types of symmetry transformations that can occur in plane figures. All symmetry transformations keep the basic figure the same size and shape without stretching, shrinking or distorting the shape of the figure.

The first transformation we will consider is *translation*. A translation moves the basic figure without turning or flipping it as shown in Figure 5.

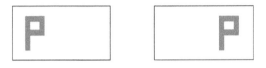

Figure 5. Translation of a single figure within the boxed area.

For a pattern to have translation symmetry, the entire pattern must match up after a translation, as in Figure 6.

Figure 6. Repetition of a figure with translation symmetry extending infinitely in two directions along a single line.

In a *rotation* a single point remains fixed while all other points move (rotate) about that point by a specific number of degrees, the angle of rotation. See Figure 7.

Figure 7. Rotation by 90° of a single figure within the boxed area about the point marked in green.

In general (although not in cross-stitch), any angle of rotation that evenly divides 360° is possible. For a pattern to have rotational symmetry, the entire pattern must match up after rotation, as in Figure 8.

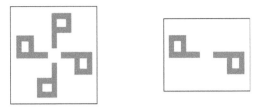

Figure 8. Two figures with rotational symmetry. The one on the left has 90° rotational symmetry, and the one on the right has 180°.

In a *reflection*, a line is fixed (see Figure 9) and acts like a mirror, reflecting each side to the other.

Figure 9. Reflection of a single figure within the boxed area.

In a pattern with reflection symmetry, the sides will match if the pattern is folded along the line of reflection as in Figure 10.

Figure 10. A figure with reflection symmetry. The line of reflection is shown in black.

The fourth and final transformation is a combination of translation and reflection and is called a *glide reflection*. As can be seen in Figure 11, the line of reflection and the line along which the pattern is translated are always parallel.

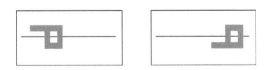

Figure 11. A glide reflection transformation with the line of reflection shown.

For a symmetry pattern to include a glide reflection, the reflected images must be equally spaced in both directions along the line of the translation. See Figure 12.

Figure 12. A pattern exhibiting glide reflection symmetry.

In cross-stitch, any symmetry transformation must move a square of the fabric onto itself or another square. One can also see that there should be a smallest repeating piece of the pattern and a set of transformations from which one can recover the entire pattern. This smallest piece is called a *motif* or in mathematical terms a *fundamental domain*.

1.2   Categories of Symmetry Patterns

Let's look at the three categories of symmetry patterns and see which ones are possible in counted cross-stitch. In rosette patterns, at least one point is fixed by any symmetry. Since translations and glide reflections do not hold any points fixed, the only transformations that can occur in rosette patterns are reflections and rotations. As delineated in Section 2.1, there are six basic rosette patterns possible in counted cross-stitch. These are shown in Figure 13.

The first image on the left in each row is the basic motif. It has no rotation or reflection symmetries. The second image shows a rosette pattern with only a single reflection symmetry. The third image shows a 180° rotational symmetry. The fourth shows two reflection symmetries, the lines (mirrors) of which are perpendicular to each other. This image also has a 180° rotational symmetry. The fifth image shows a pattern with 90° rotational symmetry but no reflection symmetries. The sixth and final image shows a pattern with 90° rotational symmetry and reflection symmetries in four different lines, all meeting at the center of the pattern with 45° between adjacent lines of reflection.

There are seven basic frieze patterns, and all of them are possible in counted cross-stitch. These are demonstrated in Figure 14 using the same motifs as in the rosette patterns. Details of the mathematics involved appear in Section 2.2.

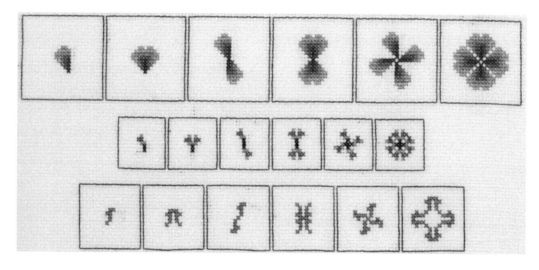

Figure 13. Three demonstrations of the possible rosette patterns in cross-stitch.

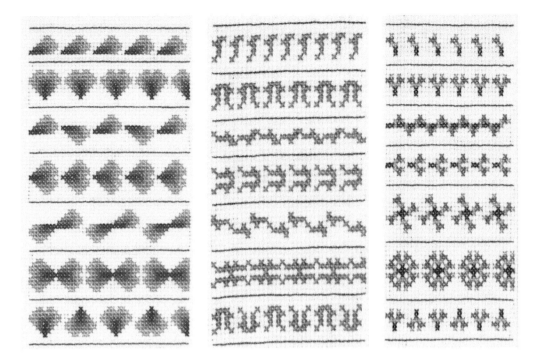

Figure 14. The seven basic frieze patterns.

In Figure 14, the top row in each image shows the basic motif translated, but with no other symmetries other than the translation. The second row shows a pattern that has a reflection in a line perpendicular to the direction of the translation, called a vertical mirror. This row includes two different mirror lines, one through the center of each motif and one between pairs of adjacent motifs. The third row shows glide reflection symmetry, but no additional symmetries besides the required translation symmetry. The fourth row shows reflection symmetry in a line that is in the same direction as the translation, called a horizontal mirror. The fifth row shows a pattern with only a 180° rotational symmetry. Notice that there are two centers of rotation: one in the middle of the motif pair (or quartet) and the other between motif pairs (or quartets). The sixth row has both horizontal and vertical mirrors and a 180° rotational symmetry, but no glide reflection. The seventh and final row shows a pattern with vertical mirrors, a glide reflection, and 180° rotational symmetries, but no horizontal mirrors.

Within cross-stitch, twelve of the seventeen wallpaper patterns are possible. Section 2.4 explains why the remaining five patterns are not possible. Figure 15 gives a demonstration of the twelve possible patterns using the half-heart motif.

2 Mathematics

The geometry of the Euclidean plane is basic to the discussion that follows.

Theorem 1 [7]

(a) *Isometries of the plane are a group under composition.*

(b) *Any isometry of the plane can be decomposed into the composition of at most three reflections.*

(c) *The composition of two reflections is either a rotation by twice the angle between the reflection lines (if the reflection lines intersect), or a translation by twice the distance between the lines of reflection (if the reflection lines are parallel)*

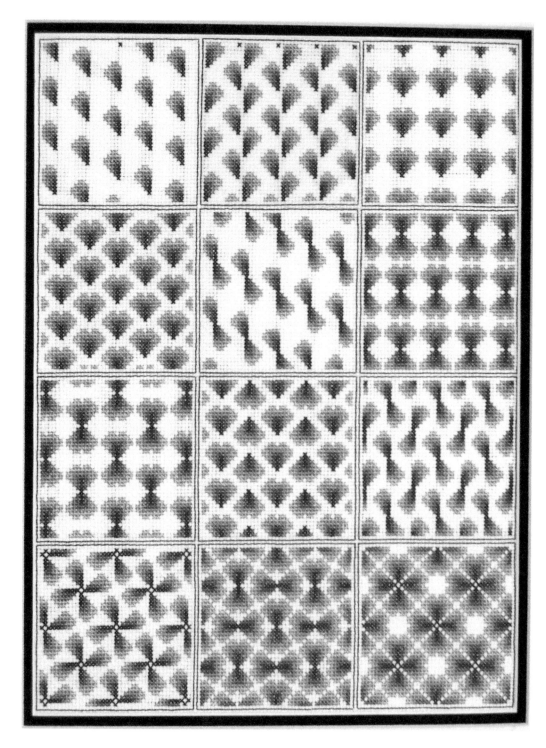

Figure 15. The twelve possible wallpaper patterns.

Cross-stitch patterns represent subgroups of the group of isometries of the plane. We will now review the classifications of rosette, frieze, and wallpaper patterns and show which are realizable by cross-stitching.

2.1 Rosette Patterns

A rosette symmetry pattern has at least one fixed point. Thus, it has no translation symmetry and no glide reflection symmetry, and only rotational and reflection symmetries are possible. Because a square of the fabric must rotate onto another square of the fabric, only rotations of multiples of 90° are possible. (This is another reason why 60° and 120° rotations are not possible in cross-stitch patterns.) Thus, a cross-stitch pattern may have 0°, 90°, or 180° rotational symmetry.

Lines of reflection must go through the fixed points. Note that if there is more than one line of reflection, this means that there can be only one fixed point, which will be the center of the pattern. The number of different reflections possible are 0, 1, 2 and 4. The lines of reflection are separated by equal angles. Thus when two lines of reflection are present, the lines meet at 90° angles, and when four lines of reflection are present, the lines meet at 45° angles.

It is possible to have a rotation without a reflection and vice versa, so we might expect there to be 12 cross-stitch rosette patterns (3 rotations times 4 reflections), but this is not the case. If there are two lines of reflection, these lines meet at 90° and by Theorem 1(c), there is then a 180° rotation. Similarly, four lines of reflection imply a 90° rotation. This reasoning proves the following theorem.

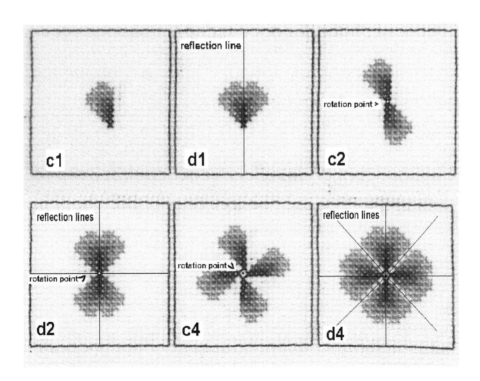

Figure 16. Rosette patterns, labeled, with lines of reflection and centers of rotation marked.

Theorem 2 *There are exactly six cross-stitch realizable rosette patterns:*

1. *c1—no rotations, no reflections;*

2. *c2—180° rotations, no reflections;*

3. *c4—90° rotations, no reflections;*

4. *d1—no rotation, one line of reflection;*

5. *d2—two lines of reflection, and thus a 180° rotation;*

6. *d4—four lines of reflection, and thus a 90° rotation.*

The standard nomenclature for symmetry patterns bears some explanation. A "c" refers to a transformation being cyclic, and the number following is the order of the transformation. The pattern c1 could be considered a motif, as it has no symmetries. A "d" stands for "dihedral" (involving reflections) and the number following is the number of distinct lines of reflection. Figure 16 gives examples of each rosette pattern in cross-stitch, complete with reflection lines and centers of rotation labeled.

2.2   Frieze Patterns

The single translation that defines a frieze pattern also limits the possible additional symmetries. For example, a frieze pattern can have 180° rotational symmetry, but not 90° rotational symmetry, as the single translation direction must be mapped onto itself. Similarly, there may be reflections, but the lines of reflection may only be perpendicular to or parallel to the translation direction (commonly called vertical and horizontal mirrors of the pattern, respectively). Furthermore, the single translation forces a glide reflection to have its line of reflection coincident with the translation.

Theorem 3   [7]   *There are exactly seven possible frieze patterns, as follows:*

1. *p111—translation symmetry only;*

2. *pm11—vertical mirrors only (there will be two of them);*

3. *p1g1—glide reflection only;*

4. *p1m1—horizontal mirror only (there will be only one);*

5. *p112—180° rotation only (there will be two centers of rotation);*

6. *pmm2—vertical and horizontal mirrors and 180° rotations (no glide reflections);*

7. *pmg2—vertical mirrors, glide reflection and 180° rotations, but no horizontal mirrors.*

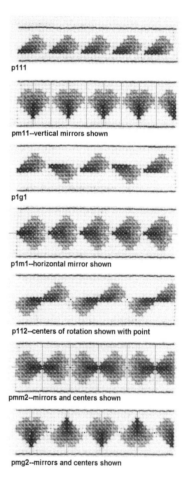

Figure 17. The seven frieze patterns, labeled, with the symmetries shown.

Our nomenclature for frieze patterns follows that adopted by the International Union of Crystallography. Each code has the form p*xyz*, where the p stands for "primitive" and denotes the type of fundamental domain present [5]. A 1 in any of the *xyz* positions indicates the absence of the corresponding symmetry. The *x* position indicates a vertical mirror by an m. The *y* position indicates the presence of a horizontal mirror by an m, or the presence of a glide reflection by a g. A 2 in the *z* position indicates the presence of 180° rotations. All seven frieze patterns are possible in cross-stitch, as shown in Figure 17.

2.3   Rosette Patterns that Generate Frieze Patterns

It is interesting to note that the translated motif for the frieze group p112 must have 180° rotational symmetry but no reflection symmetries. The rosette patterns c2 and c4 have this property and either can be used as a basis for p112. Similarly, frieze pattern pmm2 requires the translated pattern to have 180° rotational symmetry and both horizontal and vertical reflections. Rosette patterns d2 and d4 meet these criteria and may be used as bases for pmm2. Even though they have 90° rotational symmetry, both c4 and d4 can be bases for frieze patterns because the 90° rotation is not "used" under translation. The third set of frieze patterns in Figure 14 uses the c4 and d4 rosette patterns; the first two sets use the c2 and d2 patterns.

2.4   Wallpaper Patterns

Wallpaper patterns have translation symmetry in two linearly independent directions. They are significantly more complicated to study than frieze patterns because of the greater variety of possible rotations and reflections.

Theorem 4   [1] *There are exactly seventeen possible wallpaper patterns.*

Corollary 1   *There are exactly twelve cross-stitch realizable wallpaper patterns as listed in Table 1.*

Proof: The only rotations possible in wallpaper patterns are 0°, 60°, 90°, 120°, and 180°. (This is called the crystallographic restriction.) Because any pattern with either a 60° or 120° rotational symmetry is not possible in cross-stitch, we can omit the wallpaper patterns that have 60° and 120° symmetries. Excluding these five patterns leaves us with twelve possible wallpaper patterns. All of these patterns are realizable in cross-stitch, as shown in Figure 15 and again in Figure 18 labeled with invariant sets and International Union of Crystallography codes.   □

Lines of reflection in cross-stitch wallpaper patterns must be vertical, horizontal or diagonal (exactly 45°) with respect to the grid of the fabric. Points of rotation can occur at corners or centers of squares. A point of rotation can be the midpoint of the side of a square, but not when a 90° rotation is present. This transformation would take a square with corners on the grid to a square with corners at the centers of four squares, and not back onto the grid of squares of the fabric. Translations can be in any rational direction, by moving horizontally and vertically along the fabric by whole grid counts. A glide reflection axis can only be horizontal, vertical or at a 45° angle to the grid of the fabric.

2.5   Additional Mathematical Considerations

The study of symmetry patterns is one of the places in mathematics where algebra and geometry come together. From an algebraic perspective, symmetry

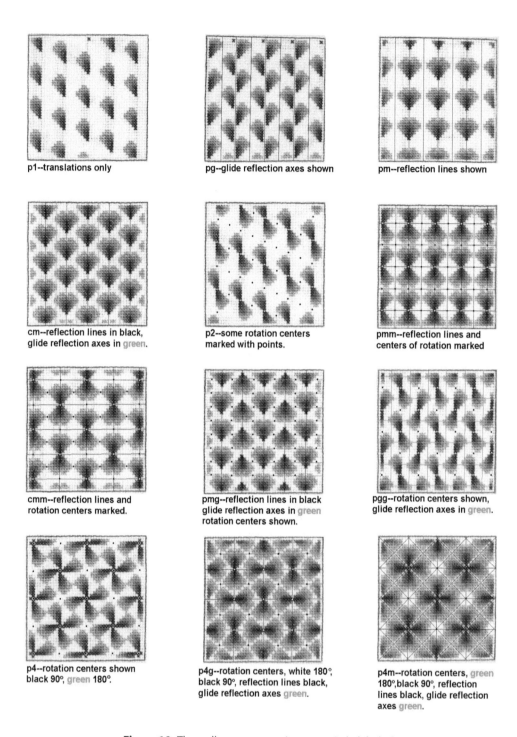

p1--translations only

pg--glide reflection axes shown

pm--reflection lines shown

cm--reflection lines in black, glide reflection axes in green.

p2--some rotation centers marked with points.

pmm--reflection lines and centers of rotation marked

cmm--reflection lines and rotation centers marked.

pmg--reflection lines in black glide reflection axes in green rotation centers shown.

pgg--rotation centers shown, glide reflection axes in green.

p4--rotation centers shown black 90°, green 180°.

p4g--rotation centers, white 180°, black 90°, reflection lines black, glide reflection axes green.

p4m--rotation centers, green 180°, black 90°, reflection lines black, glide reflection axes green.

Figure 18. The wallpaper groups in cross-stitch, labeled.

Code	Rotation	Reflection	Glide Reflection	Notes
p1	none	No	No	–
pg	none	No	Yes	–
pm	none	Yes	–	Glide reflection axis along line of reflection.
cm	none	Yes	–	Glide reflection axis not along line of reflection.
p2	180°	No	No	–
pgg	180°	No	Yes	–
pmg	180°	Yes	–	All reflections parallel.
cmm	180°	Yes	–	Non-parallel reflections, some rotation centers not on reflection axes.
pmm	180°	Yes	–	Non-parallel reflections, all rotation centers on reflection axes.
p4	90°	No	–	–
p4g	90°	Yes	–	Lines of reflection do not intersect at 45°.
p4m	90°	Yes	–	Lines of reflection intersect at 45°.

Table 1. The twelve wallpaper patterns realizable in cross-stitch, coded using International Union of Crystallography nomenclature.

patterns represent groups of isometries of the plane. The rosette patterns give a nice visualization of the group D_4 (the symmetries of a square) and its subgroups. The frieze and wallpaper patterns give visual representations of discrete but infinite groups. In the geometric view, symmetry patterns correspond to orbifolds, which were developed by Bill Thurston. An *orbifold* is the quotient of a manifold by a discrete group acting on it ("orbifold" = "orbit" + "manifold"). J. H. Conway developed an orbifold notation for the frieze and wallpaper patterns [2]. Two websites with more information about orbifold and crystallography notation are [8] and [6].

Perhaps because of the visual nature of symmetry patterns, there are many sources that give examples of the various symmetry groups. Xah Lee's website [8] is particularly useful and has a good list of references. Donald Crowe's paper [3] has a variety of images of wallpaper patterns, and the first part of Michael

Shepperd's [6] website has a nice exposition on frieze patterns.

3 Teaching Ideas

Identifying symmetry patterns can be done in elementary school. Using several sets of rosette patterns on coasters, a teacher might ask young students to sort them into groups of patterns with reflection symmetries or those with only rotational symmetries. Frieze patterns can be handled in the same way with rulers that have the patterns attached. Middle-school students can create symmetry patterns on geoboards or on graph paper, and cross-stitch ornaments or coasters with their designs. They might explore whether or not other symmetries are possible for the translated motif in a frieze or wallpaper pattern.

At the high-school level, students can use cross-stitch patterns to see that $\tan(60°)$ is irrational. They can set calculators to degree mode and verify that the decimal tangent of 60° does not appear to end or repeat.

College students in modern algebra can use symmetry patterns to explore some groups and subgroups via geometry. Students might be asked to explore or explain how the six different rosette patterns exhibit the subgroups of the group D_4. Later in the term, they can be asked to find a way to exhibit the cosets of some of the subgroups using cross-stitch. Similarly, students could be asked to find or demonstrate subgroups of the infinite symmetry groups represented by the frieze and wallpaper patterns. This would require additional work on the part of the students, as the images used in this chapter are not designed to illustrate subgroups clearly.

University students in a liberal arts mathematics course might be asked to identify the symmetries present in various designs. The original presentations of the symmetry patterns in Figures 13–15 or the project presented at the end of this chapter could be used for this purpose. Some possible questions might include the following:

★ Which symmetry patterns are present?

★ Identify the axes of the reflections and/or glide reflections.

★ Identify the centers of rotation and classify the amount of rotational symmetry present at each.

★ In the chapter project, two of the patterns in the larger squares have the same symmetry pattern. Find them and identify the differences between them.

Figure 19. The Symmetries Sampler.

Warning to instructors. It is not at all easy to find all centers of rotation, axes of glide reflections, or even lines of reflection in some patterns.

Designing wallpapers or borders or rosettes can be powerful methods for understanding the groups and the symmetries involved. This could be done by students in a mathematics for elementary teachers course, liberal arts mathematics course, geometry course, or algebra course.

In addition, symmetry patterns arise in ethnographic studies. A good reference in this area is [4].

4 How to Make the Symmetries Sampler

This project, shown in Figure 19, is 118 stitches square, covering a 8.43″ (21.41 cm) square of fabric.

Materials

* A 14″ square of 14-count Aida cloth. (You may need to buy a larger piece and cut it down to size.) To prevent the fabric from fraying at the edges, wrap the edges with narrow masking tape or whip stitch around the edges.

* A #24 tapestry needle.

* A 6″ embroidery hoop, to keep the fabric tight as one stitches.

* One skein each of DMC floss in colors 304, 316, 327, 742, 777, 844, 964, 977, 987, 3726, 3812, 3855. Use 2 of the 6 strands for stitching and work with 18″ lengths of floss. One way to organize unused floss is to insert floss into 2″ slits cut into a strip of cardboard, with each slit labeled with a color name.

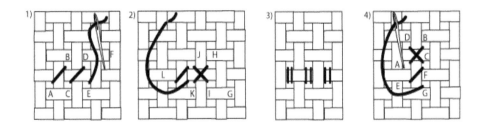

Figure 20. (1) Work all crosses through the holes of the fabric. Begin at the left and work to the right by coming up at A, going down at B, up at C, down at D, etc., across the row. (2) Complete the other half of each cross by stitching from right to left across the row, making sure all stitches cross in the same direction. (3) The back of the fabric. (4) Work vertical stitches one at a time.

Symbol	DMC No.	Color Description	Symbol	DMC No.	Color Description
●	304	MD Red	N	964	LT Seagreen
◇	316	MD Antique Mauve	⊔	977	LT Golden Brown
3	327	DK Violet	◔	987	DK Forest Green
→	742	LT Tangerine	⋈	3726	DK Antique Mauve
▼	777	V DK Raspberry	L	3812	V DK Seagreen
■	844	Ultra DK Beaver Gray	<	3855	LT Autumn Gold

Figure 21. The key.

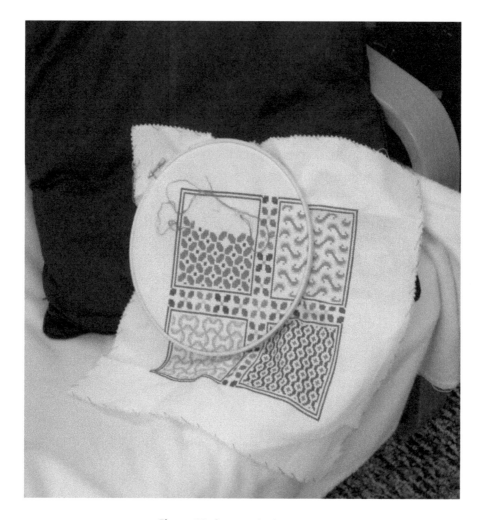

Figure 22. Symmetries in process

★ (Optional, to make a pillow) 3/8 yard of medium-weight backing fabric, such as a broadcloth.

★ (Optional, to make a pillow) Polyester fiberfill or a 12″ pillow form.

Basic Cross-Stitch Instructions

Figure 20 shows the process of making cross-stitches. To begin, bring needle up at the start of the first stitch leaving 1 inch as a tail on the back side. Hold the tail in the direction you will be stitching, and cover it with the first few stitches to hold it in place. New threads can be started by passing the needle through several stitches on the back side of the embroidery. After you have worked an area, end your thread by weaving it through the wrong side of your stitches.

Trim the loose thread ends neatly. (If you have ever seen the work of an experienced cross-stitcher, the back is almost as neat as the front—there are no knots or loose thread tails.)

Sampler Instructions

Figures 23–26 show the charts for the Symmetries Sampler, with the center of the pattern marked with

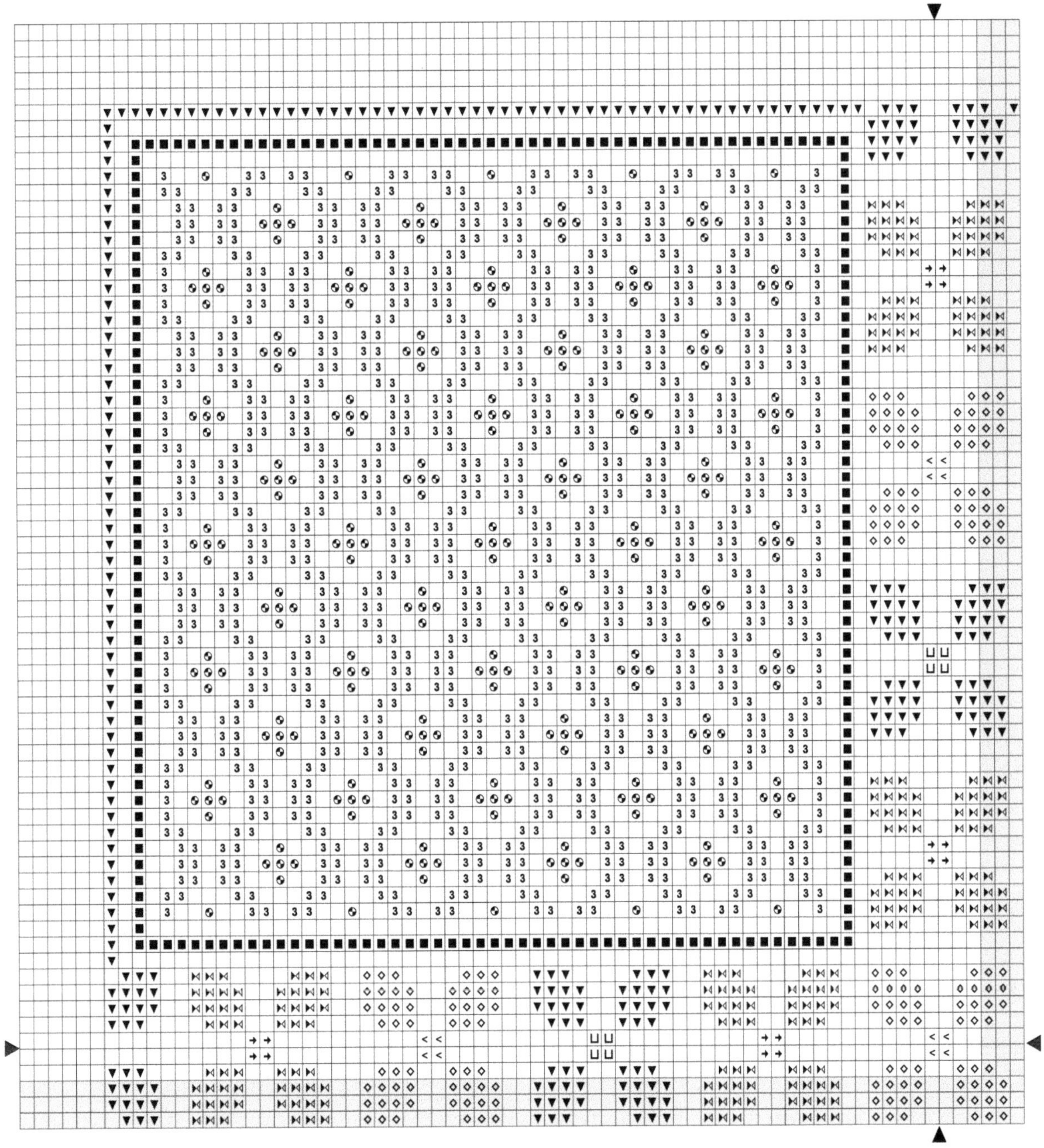

Figure 23. Upper-left quadrant.

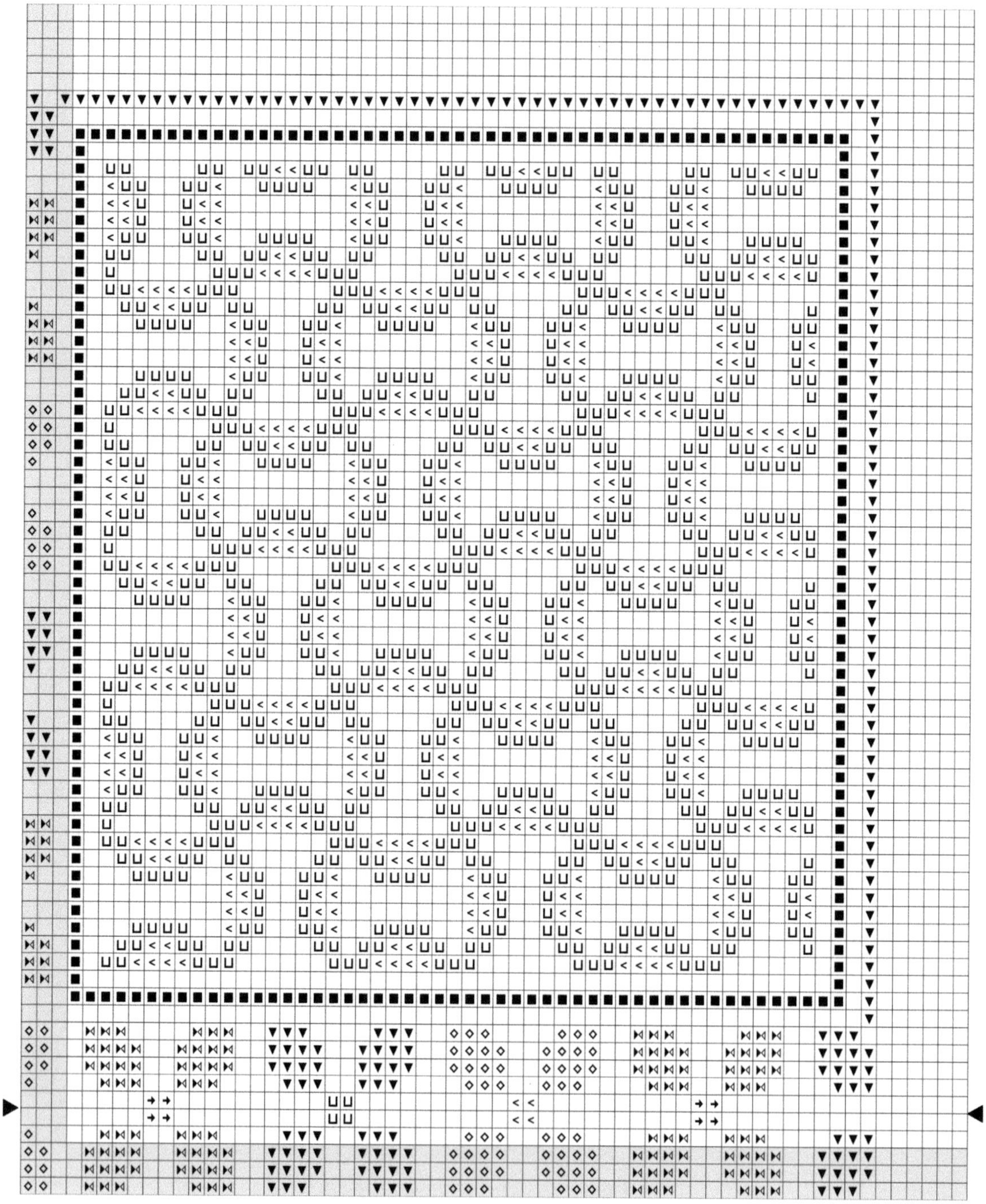

Figure 24. Upper-right quadrant.

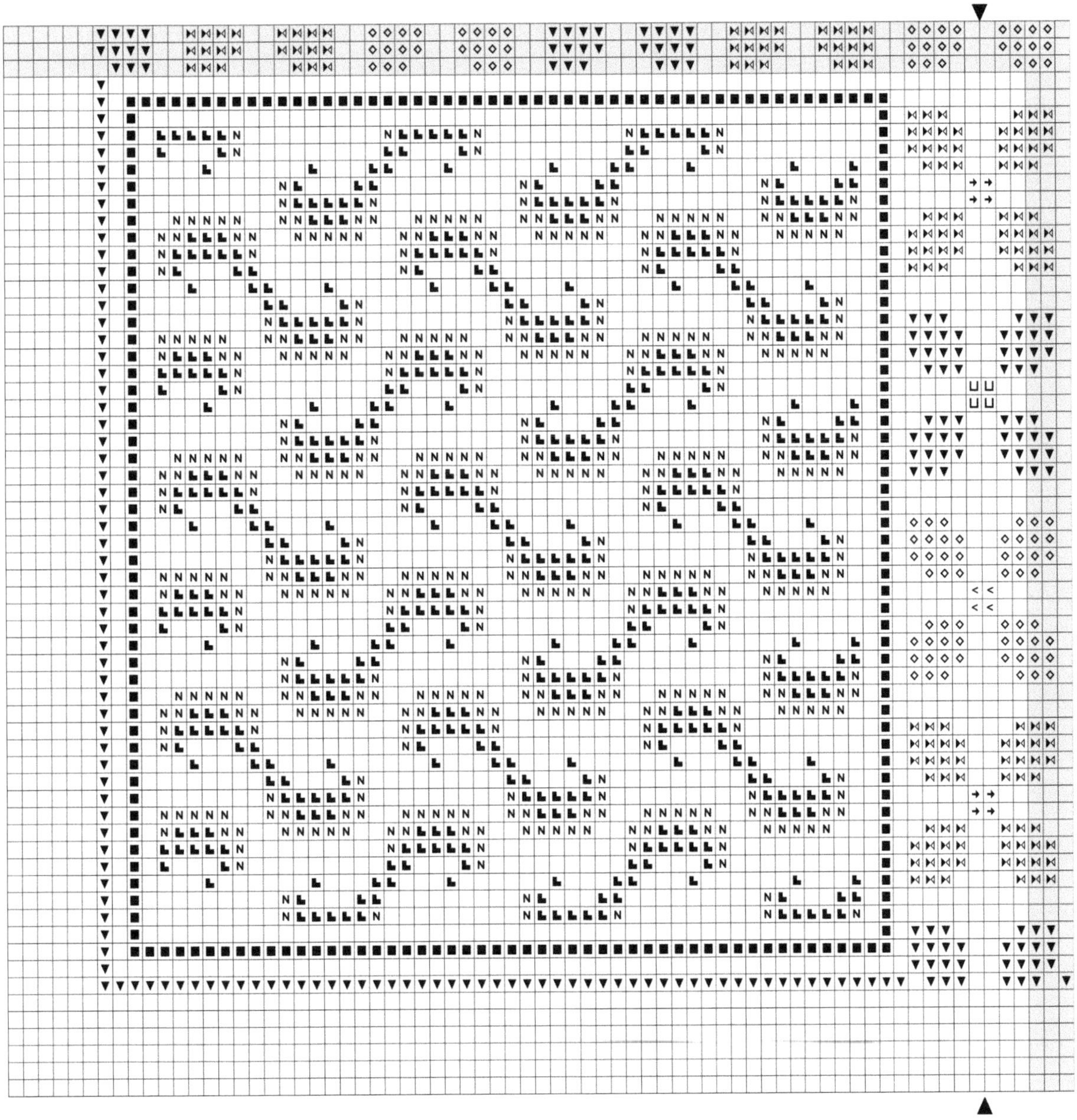

Figure 25. Lower-left quadrant.

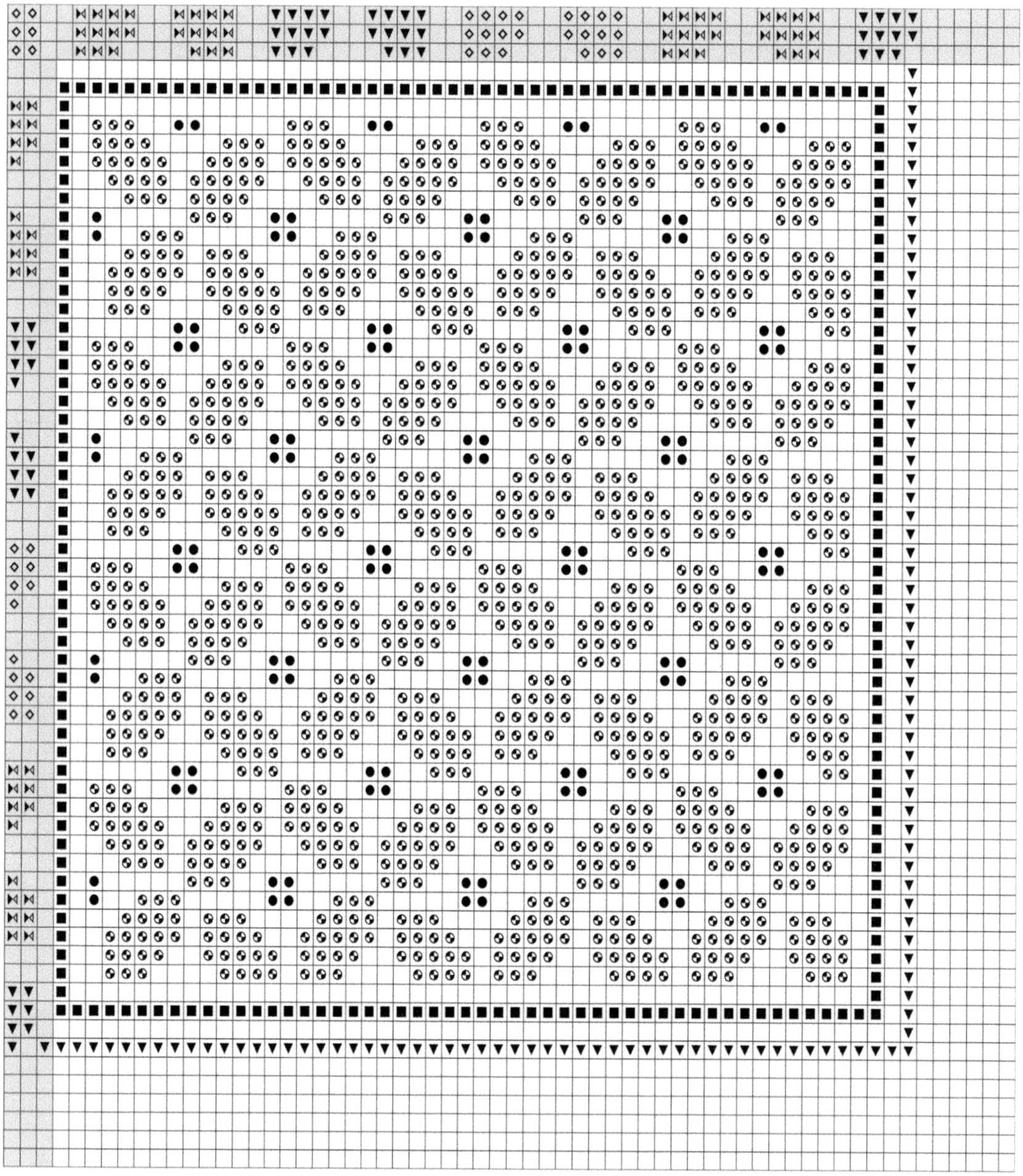

Figure 26. Lower-right quadrant.

arrows and overlap between charts shaded in turquoise. It is usually easiest to start the design at the center and work out to the edges. (To find the center, fold your fabric in half vertically, then horizontally. Where the folds meet is the center point.) Mark the center with a pin or removable piece of thread.

Cleaning and Pressing

Use a mild dishwashing liquid in lukewarm water to hand wash your design. Cold water washes made for fine washables and wool garments may actually cause the thread to bleed. Rinse in lukewarm water until the water is clear. Roll the design in a towel and squeeze well to remove excess water. Lay your design flat to air dry. To press your design after it has dried, lay it face down on a towel. Use a dry iron on a medium temperature setting and press lightly so you do not flatten your stitches.

Finishing

The completed piece may be finished as a pillow or matted and framed. (An incomplete piece is shown in Figure 22.) Consult a needlecraft or frame shop for professional framing. To complete as a pillow, trim the needlework and backing fabric to 13″ square. Sew the backing fabric to the needlework with a half-inch seam allowance. Be sure to seam along one of the lines of threads in the Aida fabric. Leave a 4″ opening at the bottom. Turn the pillow right-side out and stuff with fiberfill. Hand-sew the opening closed.

Bibliography

[1] Burn, R. P. *Groups: A Path to Geometry.* Cambridge University Press, Cambridge, UK, 1985.

[2] Conway, J. H. "The Orbifold Notation for Surface Groups." In *Groups, Combinatorics and Geometry*, LMS Lecture Notes 165, edited by M. W. Liebeck and J. Saxl, pp. 438–447. Cambridge University Press, Cambridge, UK, 1992.

[3] Crowe, Donald. "Symmetries of Culture." In *Bridges Proceedings 2001*, edited by Reza Sarhangi and Slavik Jablan, pp. 1–20. Tarquin publications, St Albans, UK, 2001. Also available at http://www.mi.sanu.ac.yu/vismath/crowe1/index.html.

[4] Crowe, Donald, and Washburn, Dorothy. *Symmetries of Culture.* University of Washington Press, Seattle, 1988.

[5] Kopsky, V., and Litvin, D. B. "Nomenclature, Symbols, and Classification of the Subperiodic Groups." *Acta Crystallographica*, vol. A49, no. 3, 1993, p. 594.

[6] Shepperd, M. "Mathematics Resources by Topic: Groups/Symmetry Patterns." http://michaelshepperd.tripod.com/resources/groups.html.

[7] Sibley, Thomas. *The Geometric Viewpoint: A Survey of Geometries.* Addison Wesley Longman, Reading, MA, 1998.

[8] Lee, Xah. "The Discontinuous Groups of Rotation and Translation in the Plane." http://xahlee.org/Wallpaper_dir/c0_WallPaper.html, 2003.

CHAPTER 6

socks with algebraic structure

CAROLYN YACKEL

1 Overview

Look at your feet. Are you wearing socks? Even if you're not, I'm willing to bet that at some time in your life, you've worn socks. Most socks are *knitted in the round*, which means that they are knitted in circular rows. The mathematical implication for analyzing this knitting is that if the rows have length 60, say, then as soon as the knitter gets to 60, the next stitch is the first stitch on the next row, or stitch number 1. That's like a clock. After 12 o'clock, we don't just keep going on to 13 o'clock (unless we're on military time), we start over at 1 o'clock. Because of this special rule, clock arithmetic works a little differently than regular arithmetic. Similarly, the arithmetic for knitting in the round works like clock arithmetic, rather than like regular arithmetic. Mathematicians call clock arithmetic *modular arithmetic*.

Now think about socks with patterns on them. When a person is in the middle of knitting a row, the most important thing is to remember what stitch in the pattern to knit next. In that case, the pattern length becomes like the number of minutes on the clock. If the pattern repeat has 10 stitches in it, then, after the 10th stitch, the pattern begins again, so the knitter is at stitch 1. That is, the arithmetic is similar to the arithmetic on a clock with only 10 minutes in each hour before the next hour starts. Mathematicians call this arithmetic modulo 10. (Knitting in the round with 60-stitch rows uses arithmetic modulo 60.) Already a knitter can see two handy places to use the notion of modular arithmetic in knitting. Alter-

natively, a mathematician can see two different applications of modular arithmetic in knitting.

In this chapter we will use modular arithmetic, along with a few other ideas from group theory, to figure out a few things about the use of patterns that are repeated around objects that are knitted in the round, like socks and sweaters. The Klein IV socks, made in Section 4, will be used as an example throughout. Sometimes mathematical generalization will be left to the reader. There will be some technical vocabulary, and some subsections will be aimed primarily at those readers with a background in group theory; however, many sections are highly accessible to all readers, and ways of thinking of the technical terms in this specific context will be given for those readers who are as yet uninitiated into the mathematics or mathematical terminology. All readers are encouraged to refer to Gallian [3].

In the mathematics section we will discuss the following issues and their resolutions.

⋆ What if the row length is not evenly divisible by the pattern length? This poses a problem because it would seem that one of the pattern repeats must be cut off in the middle (see Figure 1).

Unlike machine printing of fabric, hand knitting allows for a lot of flexibility, so that the amount of blank space between the lively portions of patterns can be varied slightly, allowing the pattern repeats to appear essentially the same, yet fit exactly around the object being constructed. The specifics of the example in the pattern sock are discussed in Section 2.1.

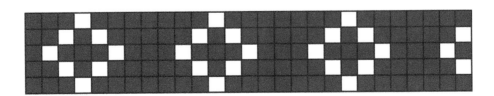

Figure 1. The row length, 27, is not evenly divisible by the pattern length, 8. Yuck!

* A different design feature appears if the pattern is (or can be made to be) only one row high. Then the pattern can be repeated past the end of a row. One convenient stopping point would be the first time the end of the pattern coincides with the end of the row. A discussion appears in Section 2.2.

* What if one wants to create diagonal stripes? Surprisingly, this problem is related to the previous problem. That is, it can be framed as a modular arithmetic problem, usually with pattern length not dividing the row length. This is done in Section 2.3.

* Next, we exploit modular arithmetic to help us avoid counting stitches when checking for errors in starting a color pattern. Section 2.4 contains an example of this technique, along with separate explanations for knitters and for mathematicians.

Important to this problem are a few ideas about knitting. First is the notion of introducing a color pattern. As carefully explained in Chapter 4, knitting involves creating rows of interlaced loops of yarn. In order to make a pattern using colors, at some point more than one color of yarn must be used when making the loops. This is called introducing a color pattern, and the knitter is often instructed by the pattern writer how to do this by using a chart. The chart is read from bottom to top, and from left to right. Samples of these charts can be seen in Figures 8–11 at the end of this chapter. The pattern is often repeated many times around an object. The knitter may be instructed to "repeat the pattern around." This instruction indicates to the knitter that each line of the chart should be repeated over and over until the end of the row, at which time the next line should be started and repeated over and over, etc. In addition, unless otherwise indicated the knitter is to assume that the end of a pattern repeat will coincide with the end of the row. Second, knitting in the round is actually knitting in a spiral rather than knitting in stacked cylindrical rows; in most discussions of knitting, however, that fact is routinely ignored. In practice, it is imperative that the knitter keep track of the first stitch in the first row so that the first stitch in each subsequent row is clear or can be ascertained. Some knitters use a stitch marker, which is a gadget that rests between two stitches, to mark the spot between the last and the first stitches of the row. Thirdly, when knitting in the round, some knitters prefer to use circular needles, which have two pointed ends joined by a cable at the back. This can be very handy, but it is not always possible for very small work (unless one manufactures a very small circular needle) because the cables tend to be long. For small work, such as socks, knitters often must turn to a set of four or five double pointed needles, which can be arranged into a polygonal shape with one free needle onto which to knit (see Figure 2). The technique described in Section 2.4 only pertains to this last situation, in which one is knitting on a set of double-pointed needles.

* Finally, Section 2.5 involves the explanation of how the well-known Klein IV group was encoded into the pattern of the sock.

In the Teaching Ideas section, we discuss a number of diverse issues one can ponder while constructing the Klein IV socks or any other pair of socks. The questions are written at a level appropriate for mathematically adept teachers to reinterpret for their students. Each item contains a line of thought that can be explored through knitting in the round. The varied topics include the effect on pieces of a pattern when 180-degree rotation is applied, factors of numbers, numbers that are relatively prime, the group theorist's notion of "order," experiencing a physical realization of $\mathbb{Z}/(60)$, and finding the mathematician's "Klein IV" group in the project sock. Although some of the topics sound like high-level mathematics, most can be restated in a way that would be

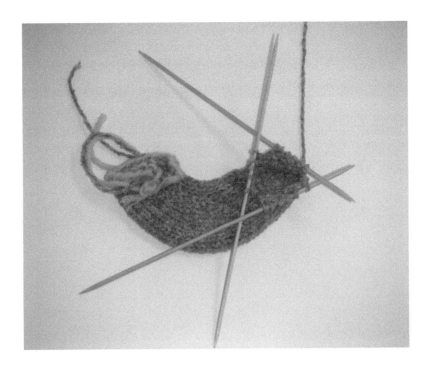

Figure 2. Double-pointed needles holding a tube.

appropriate for high school or even younger students to investigate. With the kinesthetic tool of knitting, the mathematics presented here becomes accessible to a large audience.

2   Mathematics

You might now wonder what makes something have algebraic structure. One definition of *algebra* is the study of the structure of number systems. Of course, not all algebraists agree on this or any other definition of their field! Modular arithmetic is one such number system. In particular, it has the operations of addition, subtraction, and multiplication that we have come to expect of our number systems, although it may not have division. Before you become too disenchanted with modular arithmetic because of this single flaw, however, notice that if we focus on the integers—the counting numbers (1, 2, 3, ...), their negatives, and zero—we also do not have division, since 3 divided by 2 is not another integer. It is a

rational number, but not an integer. The difficulty with division in modular arithmetic for certain moduli (clock sizes) arises from a different source, but we need not worry about that now. (Mathematicians, see [3, Chapter 25].) For now, just understand that our socks will be an excellent place to explore the structure of this new number system: modular arithmetic in the integers modulo 60 on the leg and in the integers modulo 54 on the foot, because we will be knitting in the round on 60 stitches for a large part of the leg and knitting in the round on 54 stitches for a large part of the foot.

As noted in Section 1, modular arithmetic arises naturally in two different ways when knitting in the round. The row length gives one modulus, let's call it ℓ_r, and the pattern length gives a second, usually shorter, modulus, which we will call ℓ_p. Before going on, those not familiar with modular arithmetic may want to look at how this takes place in terms of the sock. In particular, we will focus on modular arithmetic where the modulus is the row length. Addition of p and q amounts to the number of

Figure 3. Algebraic socks in action.

stitches past the starting point of the last row, obtained by knitting p stitches and then q stitches or vice versa. Multiplication of p and q is the number of stitches past the starting point of the last row obtained by knitting p stitches q times. Alternatively, $p \cdot q = q \cdot p$ is the number of stitches past the starting point of the last row obtained by knitting q stitches p times. As we know from commutativity of multiplication (the fact that the order of the terms doesn't matter) in the integers, the two are equal. Of course, $p \cdot q$ modulo n (or $\mod n$) may be less than $p \cdot q$ in $\mathbb{Z}$, the integers. Actually, really amazing things happen with multiplication in $\mathbb{Z}/(n)$, as it's called. We can have $p \cdot q \equiv 0 \mod n$, when neither p nor q is 0. (The $\equiv$ sign is read as "equivalent" or "congruent" and is just a fancy way of saying "the same" under particular circumstances.)

2.1 Varying Pattern Length

Pattern length can vary and need not divide row length. Recall that if m is evenly divisible by n, we say that n *divides* m and write $n|m$. One might assume that ℓ_p would always divide ℓ_r so that a certain number of copies of the pattern would be repeated around, but this is not always the case, as will be seen with the project sock presented. This sock begins with 72 stitches and 9 copies of a pattern of length 8; however, over the course of the leg, the number of stitches is decreased to 60. (See Figure 8 in Section 4.) In fact, there is a decrease of 3 stitches per row after each diamond. This necessitates that not every diamond pattern continue to have 8 stitches. Indeed, in the second layer of diamonds, 3 diamonds have pattern length of 7 stitches and the other 6 have 8. Further, in the third layer of diamonds, 6 diamonds have 7 stitches and the other 3 have 8. Finally, in the fourth layer of diamonds, all of the diamonds have only 7 stitches in their pattern length.

The situation we have just described is anomalous, but distinguishes a carefully hand-knitted garment from a machine-knitted one. If the pattern length varies, then modular arithmetic is not helpful. A modulus is something that should be coming up over and over as a section length (pattern length, row length, etc.), and the varying pattern length prevents this from occurring. However, the anomaly is temporary and is caused by a needed alteration in the pattern; we had to reduce the row length and thus adjust ℓ_p to compensate. Notice that we began with ℓ_p dividing ℓ_r and ended with ℓ_p not dividing ℓ_r, because no number close to 8 divides 60 (using either 6 or 10 would destroy the original pattern). In essence, the fact that ℓ_p need not be fixed (or divide ℓ_r) allows us to place patterns on shaped garments in ways so that the pattern appears to repeat evenly around the garment, even when that is not the case.

2.2   Fixed Pattern Length

Pattern length can stay fixed although it does not divide row length. For those who don't know group theory: Suppose a pattern begins at a specific stitch on a row. The number of pattern repeats that must be made before the pattern again begins on that stitch in a subsequent row is what group theorists call the order of ℓ_p modulo ℓ_r. The shorthand notation for ℓ_p modulo ℓ_r is $\text{ord}_{\ell_r}(\ell_p)$. Of course, group theorist readers now must verify that this is actually the order of ℓ_p mod ℓ_r.

Let us consider a different situation, in which the pattern length does not vary, but the pattern length still does not divide the row length. To do this, look at Figure 9 in Section 4. At this stage in the creation of the

sock, the row length is 60 and the pattern length is 24. Because the order of 24 mod 60 is 5, there will be 5 pattern repeats before the pattern will begin at the beginning of a row again. Indeed, because $24 \times 5 = 120$ and $120 \div 60 = 2$, these 5 pattern repeats take up 2 rows. In general, given a pattern of length ℓ_p and a row of length ℓ_r, the number of rows ℓ_p will repeat around before starting at the same place again is $(\text{ord}_{\ell_r}(\ell_p)) \cdot \ell_p \div \ell_r$. The reader experienced in group theory will note that this quantity is always an integer. Other readers will realize from context that if the formula is correct, it must be an integer.

Notice that when ℓ_p does not divide ℓ_r, the pattern does not "line up" from one row to the next. At the most basic level, this means that the pattern will not produce vertical stripes. More interestingly, it means that the designer has an opportunity to create a pattern that looks complex but is simple to execute. That is, instead of giving a different pattern for each row (as in Pattern A), only one instruction is needed for several rows of circular knitting. This idea leads us to the next section.

2.3   Creating Diagonals

In the previous situation, our example of $\ell_p = 24$ and $\ell_r = 60$ did not consist of relatively prime numbers (those with no common factor other than one). Although that was not necessary for the discussion of order, let's turn our attention from pattern lengths to pattern types. Suppose we wanted to have diagonal stripes advancing across the sock, as in Figure 4. We could achieve this by having a pattern length of 61, with a pattern consisting of a block of navy followed by a block of white. The shift creating the advance arises because $61 \equiv 1 \mod 60$. Again, here our pattern length does not divide our row length, but this time $\ell_p > \ell_r$. To have the diagonal going the other direction, use a pattern length of 59. Notice that to create a diagonal, merely requiring that ℓ_p and ℓ_r be relatively prime is not sufficient.

The crucial issue becomes the value of $\ell_r \mod \ell_p$. This indicates the horizontal shift of the pattern if repeated without regard to the end of the row. To create diagonal stripes, the pattern shift should cause white blocks to land underneath white blocks of the previous row, and the ends of the blocks should not coincide.

Figure 4. A pattern with diagonal stripes.

2.4 Checking for Errors without Counting Stitches

Next let's consider cases in which $\ell_p < \ell_r$ and ℓ_p divides ℓ_r.

Exploiting these common cases allows us to solve the problem of checking for errors in starting a color pattern without counting stitches. We now expect pattern length to stay constant across a single row. We note that this is sufficient for ℓ_p to act as modulus in a meaningful way, as shown in the solution to the following problem. When first beginning a color pattern, as at the top of the diamonds in the project sock, it is easy to miscount the stitches and get "off-pattern." This is not as easy later on when the knitter can use both counting and placement relative to colored stitches in previously made rows to ascertain color location. In fact, many knitters follow the pattern relative to the previous row rather than by counting, which makes patterns appear correct in the final product, even if their placement is not. For these knitters in particular, correct placement of the initial col-

ored stitches becomes even more imperative. Counting and recounting stitch placement is less than thrilling, and a more interesting solution to this problem involves using the pattern length and the number of stitches per needle. Unfortunately, this technique only works if the knitter is using a set of double pointed needles. In addition, this method won't catch all mistakes, it will only tell us if we are off!

For Knitters.   Suppose we are part-way through knitting the first row of the color pattern. We want to know if we've made a mistake. The idea is to count as little as possible, so we determine what needle we are on and to which end we are closest. If we are on the second needle, for example, we assume that the number of stitches on the first needle is correct. If we are near the beginning of the second needle, we count the stitches from the beginning of the second needle. If we are near the end of the second needle, however, we count the number of stitches to the end of the second needle and also assume that the number of stitches on the second needle is correct. This explains why we must be using double-pointed needles. Otherwise we do not have the work broken up into sections for us, and we have to count to one end or the other of the work. Our goal is to determine what stitch number we are on. So if we are at the beginning of the second needle, we add the number of stitches on the first needle to the number of stitches we've already done on the second needle. If we are almost finished with the second needle, we add the number of stitches known to be on the first and second needles and then subtract the number of stitches we haven't yet knitted on the second needle. In either case, call our stitch number s.

Next we have to figure out what stitch that should be in terms of the pattern. Of course, we want to look for a distinctive stitch. For example, in the first row of the diamond pattern of the project sock, the white stitch is distinctive. Notice that this stitch is essentially the 4th stitch in a pattern repeat of 8, but let's pretend it is the kth stitch in a pattern repeat of length ℓ_p. Then if we had done the pattern correctly, we could count off groups of length ℓ_p from the beginning up to our s, except we might have some leftover stitches that didn't fit into a group of length ℓ_p. Those are the ones we should have knitted from the next pattern repeat. So if there are k stitches left over, that is, if s divided by ℓ_p has remainder k—in math terms, s is equivalent to $k \mod \ell_p$—then the last stitch knitted should be one of the distinctive stitches. If this isn't the case, then we have made a mistake and need to go back and look for it. Continuing this way, let $\bar{s}$ be the remainder when s is divided by ℓ_p. If $\bar{s} > k$, then the distinctive stitch should be $\bar{s} - k$ stitches prior. Likewise, if $\bar{s} < k$, then the distinctive stitch should be $\bar{s} + n - k$ stitches prior. If these fail to be true, then a mistake has been made.

Note that mistakes can be made that won't be found by this method. Since what we are checking is that the ending stitch is in the correct position, if we make the mistake of putting in one too many navy stitches before a white stitch and then later put in one too few navy stitches before a white stitch, those mistakes will not be found with this technique.

For Mathematicians. Let's suppose we are starting a pattern of length ℓ_p and that the kth stitch is distinctive. In practice it is essential to identify a distinctive stitch, for otherwise it is later impossible to discern where that stitch lies in the pattern. For example, in the first row of the diamond pattern of the project sock, the white stitch is distinctive. Notice that this stitch is essentially the 4th stitch in a pattern repeat of 8. Next, suppose we have knitted almost to the end of the second needle. To decide if we are still "on" the pattern, we add the number of stitches we know to be on the first and second needles and subtract the number of stitches we have not yet knit of the second needle. This tells us what stitch number, s, we are on in terms of the round, or row. Next,

we divide this number, s, by ℓ_p and take the remainder, $\bar{s}$. Our answer, $s \equiv \bar{s} \mod \ell_p$ is what stitch we should be on in terms of the pattern. If $\bar{s} = k$, we should have just made the distinctive stitch. If $\bar{s} > k$, then we should have made the distinctive stitch $\bar{s} - k$ stitches ago. If $\bar{s} < k$, then we should have made the distinctive stitch $\bar{s} + \ell_p - k$ stitches ago. If we made the distinctive stitch when we should have, then we are "on" the pattern. Otherwise we got "off" somewhere, and must use other techniques to locate our mistake. Clearly, this method will not catch two mistakes that "cancel each other out," such as having one too many navy stitches before a white stitch in one pattern repeat and then having one too few navy stitches before a white stitch in a later pattern repeat. It will, however, catch two additive mistakes, such as having too many navy stitches before a white stitch twice with no corrective mistakes.

2.5 Decoding the Klein IV Group

This section is for group theory enthusiasts.

So far the mathematics I have described has been intrinsically created by the very existence of a pattern. The final piece of mathematics was put in simply for fun, and is not a consequence of knitting, but of the pattern itself. In the section of diamonds, each column of four diamonds encodes the Klein IV group. (See Figure 8.) To mentally link the group with the figure, think of the outside white diamond as a border. Looking only at the inside crosses as group elements, use the following addition scheme: white plus white is navy, navy plus navy is navy, and white plus navy is white. Now it is easy to check that the first (top in the sock, bottom in the pattern chart) diamond interior is idempotent. The idempotent added to any other element is that element, showing that the idempotent is the identity. Every other interior, when added to itself, yields the identity. Finally, the sum of any two non-idempotents is the third non-idempotent.

Figure 5. Algebraic socks in wool/cotton.

3 Teaching Ideas

Most mathematicians, including non-knitters, immediately appreciate how naturally the integers modulo n, denoted $\mathbb{Z}/(n)$, arise in the context of knitting in the round. Therefore, knitting is an excellent venue to introduce this alternative number system through a tactile, less abstract, slower, or more exploratory introduction to the concepts. Appropriate forums could include math for liberal arts courses, math clubs, and high-school or even junior-high math teams. Truly, there is no lower age limit, although some of the questions may have to be restated or reworked to make them age- and audience-appropriate. One of the beauties of this subject is that even for a fairly quick knitter, crafting is slow going, and there is plenty of time during the repetitive work for some deep, leisurely thought about what is happening with the patterns. I begin with some questions for the students based on the patterns in Section 4, and move on from there. The careful reader will notice that some of the answers are given in the mathematics section. These questions are to be investigated. They are not asked with the idea that the student will immediately "know" the answer. That would be boring and would not utilize knitting!

Experienced teachers will note that the vagueness of these questions will make many students (and some teachers!) uneasy. For example, they are impossible to grade, and what sorts of explanations are satisfactory? Therefore, they must be used judiciously and with forethought. Furthermore, the "answers" are not provided. Many of the questions should be considered merely as fodder for productive investigation.

⋆ Pattern C (see Figure 10) has 180-degree rotational symmetry. About what point must you rotate to get this symmetry? After rotation, all the line segments have the same slope as before. Why is this? Does this always happen with lines? With odd functions? With vertical shifts of odd functions? With horizontal shifts of odd functions? Why? Is the algebraic answer satisfactory or do you want a geometric one?

Figure 6. A pattern of length 3 above one of length 2.

⋆ If you had one pattern of length 3 with three alternating colors (red, green, and blue) and another pattern starting just below it of length 2 with two alternating colors (yellow and black, see Figure 6) would you get all six possible color combinations (red above yellow, red above black, green above yellow, green above black, blue above yellow, and blue above black), or would you only get some of them? What if you had k_1 colors on the first row and k_2 colors on the second row? Assume that k_1 and k_2 divide ℓ_r. Or even assume that $\ell_r = k_1 \cdot k_2$. Note to teachers: This problem gives a hands-on understanding of relative primeness and greatest common divisor.

⋆ Let us agree that in this sock we work in $\mathbb{Z}/(n)$ for at least some values of n. For which values of n do you think this is the case? Explain.

⋆ Why do you think that the instructions say to repeat Pattern B (see Figure 9) five times?

⋆ Make a chart for four rows of Pattern B. Can you design a similar pattern (one row, of length not dividing 60) so that the corresponding chart shows geometrical shapes?

⋆ Once the knitter gets to the part of the sock where there are 60 stitches in a row, three patterns are used: B, which has 24 stitches in a repeat, C, which has 15 stitches in a repeat, and D (see Figure 11), which has 4 stitches in a repeat. Do all of these pattern repeat numbers seem intuitive? Why or why not? Are there other pattern repeat numbers that you might have expected? Why or why not? Discuss this. Name all numbers you might have expected, if there are any others. If you were making the pattern, what pattern repeat lengths would you have used?

⋆ For each of the pattern repeat lengths mentioned in the previous question consider the following: If you started the pattern at the beginning of the row, how many pattern repeats would you need to make before the beginning of the pattern would be at the beginning of the row again? (This is called *order* to group theorists.) Can you conjecture what this number is even when your pattern repeat doesn't divide 60? Experiment by knitting.

⋆ Suppose you want to make diagonal stripes as in Section 2.3. How could you change the slope of the stripes to make them more steep or less steep?

⋆ Describe where and how the Klein IV group is displayed visually in this sock.

Figure 7. The algebraic socks on happy feet.

★ Discuss ring properties with your class and ask them to investigate these properties in terms of $\mathbb{Z}/(60)$ and the sock. Recall that this will be slightly tricky since the sock version of $\mathbb{Z}/(60)$ is $\{1, 2, \ldots, 60\}$ rather than $\{0, 1, \ldots, 59\}$.

4 How to Make Algebraic Socks

This pattern was inspired by [1] and assisted by [2].

Materials

★ One set of 4 double-pointed size 2 bamboo needles, and perhaps others if knitting for a particularly shapely calf.

★ Eight needle caps or a bunch of rubber bands, so that the needles don't fall out when you carry around your project.

★ 5 50g balls of Valley Yarns Superwash, 4 in Classic Navy and 1 in Natural.

Instructions

Note on gauge/sizing

This makes one pair of socks to fit a woman with a 12″ calf, using size 2 bamboo needles throughout. Find the circumference of your relaxed calf at its widest point. If it is significantly different from 12″, you will want to adjust the size of the sock. Take your calf measurement in inches and divide by 12 to get a fraction f. Because the number of stitches is so important, I suggest swatching and then changing needle sizes to obtain gauge rather than adjusting the number of stitches in the pattern. Your goal is to change needle sizes so that your horizontal gauge is $6f$. If you have slender ankles or want very form-fitting socks, you may wish to switch to smaller-size needles beginning with Row 44.

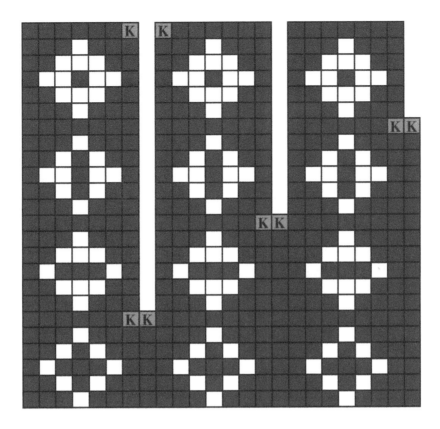

Figure 8. Pattern A.

Definition of SSK

Slip one stitch as if to knit it. Slip another stitch as if to knit it, then slip these two stitches back onto the first needle without changing their orientation. Now knit them together. The first two moves twist the stitches. The second two moves adjoin them.

Start with the Leg

In navy, cast on 72 stitches. Distribute the stitches equally on 3 needles.

Rows 1–12: Work *Knit 2, Purl 2, repeat from * around for 12 rows to create the ribbing.

Row 13: Work 1 row all Knit in navy.

Rows 14–37: Work pattern A. Recall that knitters always read charts from the bottom row to the top row. Note also that two orange Ks next to each other indicates knitting those two stitches together knitwise. This occurs in rows 19, 25, 31, and 37, which explains the subsequent "holes" in the patterns.

Rows 38–41: Work these 4 rows all Knit in navy.

Rows 42–43: Work pattern B 5 times. (Note: When you finish row 42, you will be in the middle of the pattern. This is intentional!)

Figure 9. Pattern B.

Rows 44–45: Work 2 rows all Knit in navy.

Rows 46–66: Work pattern C. Pattern C fits around the sock 4 times. The contrast color in each pattern repeat should be started with a different strand of yarn so that the yarn does not need to be carried behind the work for great lengths. Each strand needs to be at most half a yard in length.

Figure 10. Pattern C.

Rows 67–73: Work pattern D, using both color strands again.

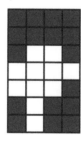

Figure 11. Pattern D.

Work the Heel Flap

The remainder of the sock is monochromatic, worked only in navy. Begin by reassigning stitches to needles as follows: Using all four needles and beginning at the beginning of the row, put 15 stitches onto each needle. Now knit the fifteen stitches from the first needle (the one containing the first 15 stitches of the row) onto the fourth needle (the one containing the last 15 stitches of the row). The fourth needle now has 30 stitches on it. This is the needle containing the stitches for the heel flap. The other two needles—needles 2 and 3—contain the stitches for the instep. Those stitches will be set aside until later. Needle 1 is free.

Work 19 rows in stockinette on the stitches on needle 4 as follows: *Slip the first stitch then purl the row. Slip the first stitch then knit the row. Repeat from * 8 more times. Slip the first stitch then purl the row. Continue working with these 30 stitches as you turn the heel.

Turn the Heel

Row 1: Knit 18, SSK (see start of instructions for the definition), knit one, and turn. (Yes, you're in the middle of a row. Don't worry about it. This will create a space that will be relevant in two rows.)

Row 2: Slip the first stitch, purl 8, purl 2 together (p2tog), purl 1, turn.

Row 3: Slip the first stitch, knit 9 (this is to within 1 of the gap each time), SSK (described above), knit 1, turn.

Row 4: Slip the first stitch, purl 10 (this is to within 1 of the gap each time), p2tog, purl 1, turn.

Row 5: Slip the first stitch, knit 11, SSK, knit 1, turn.

Row 6: Slip the first stitch, purl 12, p2tog, purl 1, turn.

Row 7: Slip the first stitch, knit 13, SSK, knit 1, turn.

Row 8: Slip the first stitch, purl 14, p2tog, purl 1, turn.

Row 9: Slip the first stitch, knit 15, SSK, knit 1, turn.

Row 10: Slip the first stitch, purl 16, p2tog, purl 1, turn.

Make the Gussets

Round 1: Slip the first stitch. Knit across the heel needle. With the same needle, pick up 11 stitches along the side of the heel flap. (For instructions, see Chapter 2, Section 1.1.) Next knit across the 30 instep stitches, 15 from needle 2 and 15 from needle 3, knitting them all onto one needle. Now pick up 11 stitches along the other side of the heel flap, and knit 10 stitches from the heel needle onto this needle.

Round 2: Needle 1: Knit 18 stitches, knit 2 together (k2tog), knit 1. Needle 2: Knit 30. Needle 3: Knit 1, k2tog, knit 18.

Round 3: Knit around.

Round 4: N 1: Knit 17, k2tog, knit 1. N 2: Knit 30. N 3: Knit 1, k2tog, knit 17.

Round 5: Knit around.

Round 6: N 1: Knit 16, k2tog, knit 1. N 2: Knit 30. N 3: Knit 1, k2tog, knit 16.

Round 7: Knit around.

Round 8: N 1: Knit 15, k2tog, knit 1. N 2: Knit 30. N 3: Knit 1, k2tog, knit 15.

Round 9: Knit around.

Round 10: N 1: Knit 14, k2tog, knit 1. N 2: Knit 30. N 3: Knit 1, k2tog, knit 14.

Round 11: Knit around.

Round 12: N 1: Knit 13, k2tog, knit 1. N 2: Knit 30. N 3: Knit 1, k2tog, knit 13.

Round 13: Knit around.

Round 14: N 1: Knit 12, k2tog, knit 1. N 2: Knit 30. N 3: Knit 1, k2tog, knit 12.

Round 15: Knit around.

Round 16: N 1: Knit 11, k2tog, knit 1. N 2: Knit 30. N 3: Knit 1, k2tog, knit 11.

Round 17: Knit around.

Round 18: N 1: Knit 10, k2tog, knit 1. N 2: Knit 30. N 3: Knit 1, k2tog, knit 10.

Now knit until the foot measures 3″ less than the desired finished length (or 25 rows before the end of the toe, if you have changed the needle size to accommodate the gauge). Then begin the toe.

Make the Toe

Round 1: *Knit 7, k2tog, repeat from * around.

Rounds 2–4: Knit.

Round 5: *Knit 6, k2tog, repeat from * around.

Rounds 6–8: Knit.

Round 9: *Knit 5, k2tog, repeat from * around.

Rounds 10–12: Knit.

Round 13: *Knit 4, k2tog, repeat from * around.

Rounds 14–16: Knit.

Round 17: *Knit 3, k2tog, repeat from * around.

Rounds 18–19: Knit.

Round 20: *Knit 2, k2tog, repeat from * around.

Rounds 21–22: Knit.

Round 23: *Knit 1, k2tog, repeat from * around.

Round 24: Knit.

Round 25: *k2tog, repeat from * around.

Cut off a 6″ tail of yarn and pull through the remaining loops. Tie off. Pull the strand inside the sock, and weave in the end.

Bibliography

[1] Bush, Nancy. *Folk Socks: The History & Techniques of Handknitted Footwear*. Interweave Press, Loveland, CO, 1994.

[2] Coats and Clark. *Learn How Book*. Coats and Clark, New York, 1959.

[3] Gallian, Joseph. *Contemporary Abstract Algebra*, fifth edition. Houghton Mifflin Company, New York, 2002.

CHAPTER 7

fortunatus's purse

SUSAN GOLDSTINE

1   Overview

Lewis Carroll is so universally recognized as a writer that most of his readers are blissfully unaware that he was also a professor of mathematics. Among his classic works that have entertained generations of children and adults are *Alice's Adventures in Wonderland*, *Through the Looking Glass and What Alice Found There*, and *The Hunting of the Snark*. Not among his classic works is the two-volume ramble *Sylvie and Bruno* and *Sylvie and Bruno Concluded*, which seems to most modern readers an overlong, overwrought, overly Victorian concoction of fairy children and moralizing. However, there are several pages of *Sylvie and Bruno Concluded* [1] that merit the attention of the mathematically inclined reader. They come at the beginning of the chapter titled "Mein Herr," referring to a mysterious, nameless German professor who arrives to take tea with several of the principal characters. Another member of the party is Lady Muriel, who is hemming handkerchiefs. Inspired by her activity, Mein Herr brings up the subject of Fortunatus's Purse, a legendary ever-full bag of money.

Mein Herr announces that Fortunatus's Purse may be easily constructed from three handkerchiefs, and Lady Muriel is sufficiently intrigued to demand instructions on the spot.

'You shall first,' said Mein Herr, possessing himself of two of the handkerchiefs, spreading one upon the other, and holding them up by two corners, 'you shall first join together these upper corners, the right to the right, the left to the left; and the opening between them shall be the *mouth* of the Purse.'

A very few stitches sufficed to carry out *this* direction. 'Now, if I sew the other three edges together,' she suggested, 'the bag is complete?'

'Not so, Miladi: the *lower* edges shall *first* be joined—ah, not so!' (as she was beginning to sew them together). 'Turn one of them over, and join the *right* lower corner of the one to the *left* lower corner of the other, and sew the lower edges together in what you would call *the wrong way*.'

'*I see!*' said Lady Muriel, as she deftly executed the order. 'And a very twisted, uncomfortable, uncanny-looking bag it makes! But the *moral* is a lovely one. Unlimited wealth can only be attained by doing things *in the wrong way*! And how are we to join up these mysterious—no, I mean *this* mysterious opening?' (twisting the thing round and round with a puzzled air). 'Yes, it *is* one opening. I thought it was *two*, at first.'

Lady Muriel's observation precipitates a lively discussion among the tea party of the curious properties of the unfinished purse, including the further insight that it has only one surface. It is also remarked that this is similar to the "puzzle of the Paper Ring," which as described is clearly a paper Möbius band. In fact, the object in Lady Muriel's lap is a cloth Möbius band with a hole for the mouth of the purse.

When Lady Muriel once again asks how the bag is to be closed up, Mein Herr instructs her as follows.

'The edge of the opening consists of *four* handkerchief edges, and you can trace it continuously, round and round the opening: down the right edge of *one* handkerchief, up the left edge of the *other*, and then down the left edge of the *one*, and up the right edge of the *other*!'

[…]

'Now, this *third* handkerchief,' Mein Herr proceeded, 'has *also* four edges, which you can trace continuously round and round: all you need do is to join its four edges to the four edges of the opening. The Purse is then complete, and its outer surface—'

'*I see!*' Lady Muriel eagerly interrupted. 'Its *outer* surface will be continuous with its *inner* surface!

But it will take time. I'll sew it up after tea.' She laid aside the bag, and resumed her cup of tea.

Having resolved not to finish, Lady Muriel asks Mein Herr why the bag is called Fortunatus's Purse, to which he replies that since the inside of the purse is the same as the outside, all of the world's wealth is inside the purse. This answer delights Lady Muriel, who responds, "I'll certainly sew the third handkerchief in—*some* time, [...] but I wo'n't take up your time by trying it now: Tell us some more wonderful things, please!"

In fact, this last remark is a subtle mathematical joke on the part of the author. As a mathematician with some practical knowledge of topology, Carroll was aware that the final step of Mein Herr's directions is impossible to complete in Euclidean three-space. The purse must intersect itself, and so any attempt to stitch all the way around the third handkerchief must be confounded, leaving a second opening in the purse.

However, Mein Herr's instructions aside, one may simply forgo the initial opening of the purse (which, since the entirety of the world is already inside the purse, is irrelevant in any case) by stitching the entire top and bottom edges of the first two handkerchiefs. The result is the charming model that we will discuss in this chapter.

2   Mathematics

We will now consider the topology of Fortunatus's Purse. In all that follows, we take *Fortunatus's Purse* to mean the closed, ideal purse. In other words, the purse does not have the mouth described in Lewis Carroll's instructions, and we are assuming that the final stitch around the third handkerchief is complete. Since completing this stitch requires at least four Euclidean dimensions, we are committing the same liberty one commits in describing the glass bottle pictured in Figure 1 as a Klein bottle rather than as a Klein bottle with a puncture.

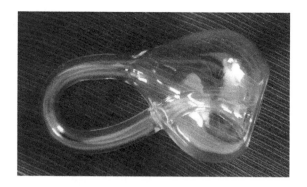

Figure 1. Klein bottle (with puncture).

With these provisos, we find that Fortunatus's Purse is that other vexing one-sided surface without boundary, the real projective plane. There are two ways to demonstrate this, each of which is enlightening in its own way.

2.1   Gluing the Square

The first method uses a mainstay technique of introductory topology, that of gluing together the opposite sides of a single square to form a smooth closed surface. In the current context, it is tempting to speak of stitching the edges rather than gluing them. However, we will be stretching the square and subsequent objects in a very non-fabric-like manner, so we stick to the traditional mathematical metaphor of gluing.

Depending on the relative orientations of the edges, we obtain three possible surfaces, shown in Figure 2. In this diagram (and many that follow), we use arrows to indicate in which direction the gluing is to take place. For instance, in the middle square the blue arrows are pointing in the same direction, indicating that the top and bottom edges are to be glued from left to right, whereas the red arrows point in opposite directions, indicating that the top of the right edge is to be glued to the bottom of the left edge and vice versa, introducing a half-twist in the surface. Gluing both pairs of edges with the same orientations yields a torus, gluing one pair with the same orientation and one pair with opposite orien-

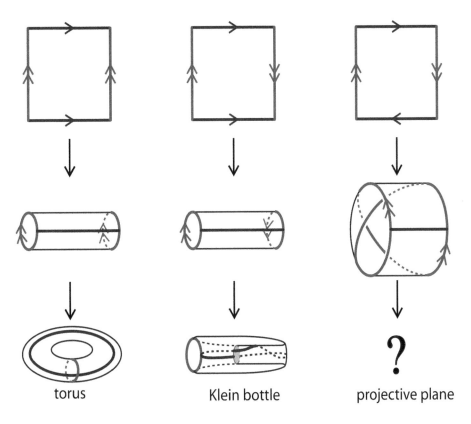

Figure 2. Three ways of gluing a square.

tations yields a Klein bottle, and gluing both pairs with opposite orientations yields a projective plane. Since the Klein bottle and the projective plane each incorporate at least one half-twist, they are both one-sided, and hence nonorientable surfaces (as in Chapter 1).

The first two constructions are fairly easy to visualize, since in each case we may start by gluing two of the edges together in the same orientation to form a cylinder. By contrast, with the projective plane we are forced to begin by making a Möbius band, and then we are left with the mind-twisting task of gluing together opposite pairs of points on the single edge of the band.

In order to make this easier to visualize, we can dissect the square before we cut it, as in Figure 3a. Although this appears as two separate diagonal cuts of the square, it is not; since the sides of the square are glued together, it forms a single circular cut in the final surface. When we assemble the pieces on either side of this cut and smooth them out (Figure 3b), we find that we are left with a Möbius band on one side of the cut and a disk on the other side of the cut. Therefore, the projective plane consists of a Möbius band and a disk with their boundaries glued together.

But this is exactly Fortunatus's Purse! For the purse, we stitch the edges of a square, which is a topological disk, to the edge of a Möbius band (Figure 3c). Since we cannot complete this final edge-stitching in the physical world, the actual piece of cloth is topologically a projective plane with a hole in it. By the logic of Figure 3 reversed, this punctured projective plane is a Möbius band, so the addition of the third handkerchief does not alter the topology of the physical model.

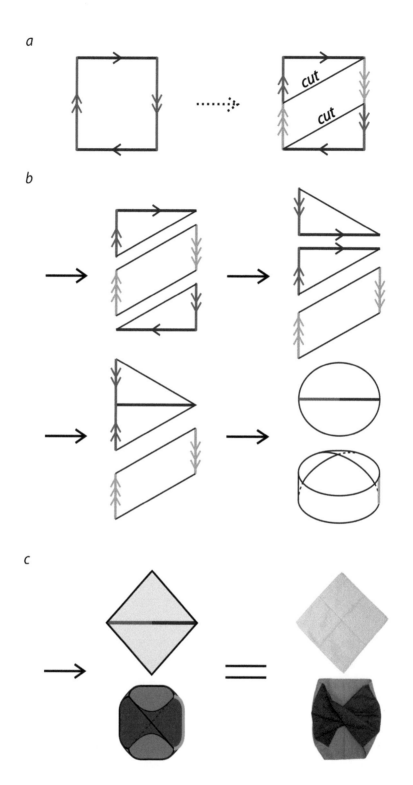

a

b

c

Figure 3. Decomposing the projective plane into the parts of Fortunatus's Purse.

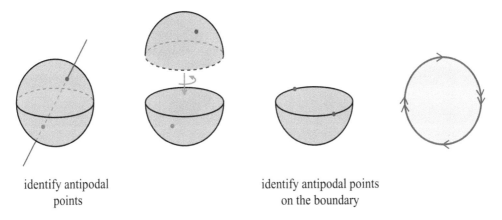

identify antipodal
points

identify antipodal points
on the boundary

Figure 4. Gluing antipodal points on the sphere.

2.2 Gluing the Cube

The second method for recognizing Fortunatus's Purse uses another standard description of the projective plane: it is the surface obtained by gluing together the pairs of antipodal points on a sphere. Again, this is hard to visualize all at once, but we can make it easier (see Figure 4) by slicing off the open upper hemisphere of the sphere. The closed lower hemisphere contains at least one point from each antipodal pair. Moreover, the entire open upper hemisphere can be glued into place by rotating it by 180° and then plunging it onto the lower open hemisphere, much like inverting a bathing cap. What remains is the closed lower hemisphere, a topological disk, in which each point of the boundary is

glued to its antipodal twin. But this is identical to taking a square, also a topological disk, and gluing the opposite edges with opposite orientations, as we did above.

On the other hand, just as a square is a topological disk, so a cube is a topological sphere. Since the cube also has antipodal symmetry, we can just as easily form the projective plane by gluing together the pairs of antipodal points on a cube. By an appropriate modification of the decomposition of the glued sphere, shown in Figure 5, we see that the end result has three square faces with their edges glued together in an appropriately twisted fashion. It should come as no surprise that the gluing pattern corresponds exactly to the stitching pattern around the three square panels of Fortunatus's Purse (Figures 6 and 13).

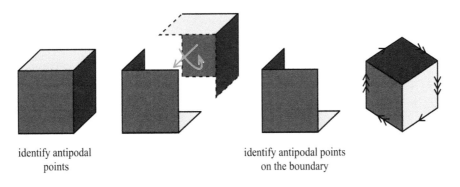

identify antipodal
points

identify antipodal points
on the boundary

Figure 5. Gluing antipodal points on the cube.

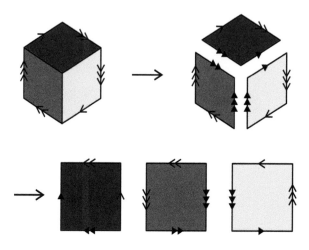

Figure 6. Cubical stitching.

This description of the purse shows that it is more symmetric than the assembly instructions suggest, for *any* pair of the three handkerchiefs forms a Möbius band, with the remaining handkerchief sewn around its edge. In the physical model, this symmetry is marred because one seam has a hole in it while the other five do not. Nonetheless, it is possible to get a sense for this symmetry by manipulating the physical purse, as described in the classroom section below.

The newfound symmetry raises an interesting question. As mentioned above and shown in Figure 13, the completed physical purse is symmetric except for the fact that one of the six sewn edges has a hole in it. But Mein Herr's instructions to Lady Muriel begin with a seam that has a hole in it. Shouldn't it be possible to pass the final seam through that initial hole to complete the purse?

The answer is that it depends on how much work you are willing to do. In order to have the final seam pass through the initial hole, you must first rip out the stitches on one side of the hole, then perform the "final" stitching, and finally resew the seam you previously undid. Topologically, it makes no difference if you rip out a seam and resew it later. This serves as a good illustration of the difference between theory and practice.

2.3 Gluing the Dodecahedron

Part of the fun of topological manipulation is that there are so many geometric ways to achieve a particular topological construction. To glue the antipodal points of a topological sphere, we may use any antipodally symmetric solid, and it is interesting to consider what happens when we begin with a regular solid other than a cube.

One of the most intriguing solids to work with is the dodecahedron. When we identify antipodal points on a solid, we halve the number of faces, edges, and vertices. From a cube, we end with three square faces joined along six edges with four vertices. From a dodecahedron, we end with six pentagonal faces joined along fifteen edges with ten vertices as in Figure 7. In fact, each pentagonal face shares exactly one edge with each of the other pentagonal faces. This corresponds to the fact that any given face of the dodecahedron touches five other faces, and these six faces together contain one from each antipodal pair.

identify antipodal points

identify antipodal points
on the boundary

Figure 7. A six-color map on the projective plane.

The result is a division of the projective plane into six regions, each bordering all the others. In other words, we have a geometric demonstration that the Four Color Map Theorem [4], which says that four colors are sufficient to color any map in the Euclidean plane so that no two neighboring countries are the same color, does not apply to maps on the projective plane! In fact, on the projective plane, we have the Six Color Theorem, though it is naturally harder to prove that six colors are sufficient than to prove that they are necessary [5, p. 67]. One interpretation of the Six Color Theorem on the projective plane appears in Figure 8.

Figure 8. A six-color projective quilt made by Jeri Riggs.

3   Teaching Ideas

Fortunatus's Purse fits very nicely into any classroom discussion of elementary topology. With a group of sufficiently outgoing students, a dramatic reading of the original *Sylvie and Bruno Concluded* scene is *de rigeur*, and for such purposes it is useful to have an extra model showing just the first two handkerchiefs stitched together as a prop. If the class has a Mathematics for Poets flavor, then Carroll's discussion of the "puzzle of the paper ring" affords an excellent opportunity to have the class make paper Möbius bands and experiment with them. It is also a marvelous place to incorporate a Möbius quilt (see Chapter 1).

In fact, one nice way to incorporate the Fortunatus's Purse episode into a liberal-arts mathematics class is to devote a portion of the course to the mathematics of Lewis Carroll. For a glimpse of how this might be done, the reader may consult [3].

After presentation of the Lewis Carroll passage, attention may turn to the properties of the model itself. The various decompositions of the model described in the mathematics section above are all classroom fodder. To make the model more fathomable (and to highlight the cubical construction from Figure 5), the purse should be made of fabric squares in three different colors. Manipulating the model, students can confirm that any two of the handkerchiefs form a Möbius band by flattening each of the three squares in turn and studying how the other two twist. Two of the squares will not flatten all the way because of the hole in the model, but they become flat enough to display the pattern. In fact, as mentioned in the instructions below, it makes practical sewing sense to use hemmed squares of fabric with discernable front and back sides, and this helps identify the Möbius twist. Whenever you flatten a "third" handkerchief and look at the four edges of the first and second handkerchiefs that are stitched to the third, you will find that one edge of each of the two handkerchiefs is a front and the other is a back. In Figure 9, the edges going counterclockwise from the upper-right are front, back, back, front.

Figure 9. Fortunatus's Purse with a Möbius band on top.

Beyond these basics, there are many directions for further classroom exploration. Here are a few of them:

* In a liberal arts math course, a nice in-class exercise is to take the written instructions from *Sylvie and Bruno Concluded* and turn them into a sewing pattern for the purse. Further challenges might include designing a sewing pattern for the dodecahedral analogue of Fortunatus's purse (as in Figures 7 and 10) to be sewn out of six pentagonal panels, as well as octahedral or icosahedral purses. Unfortunately, the sewing itself is sufficiently time-consuming that making actual models is not feasible for a class, although a student with access to a non-raveling fabric such as fleece and a sewing maching with a zig-zag stitch might undertake any of these as an outside project.

Figure 10. The Dodecahedral Challenge Purse.

* In any course that covers the Four Color Theorem, a discussion of the Six Color Theorem makes a nice extension of the topic, as does the fact that seven colors are necessary to color any map on the torus. If the class is a more rigorous course in topology, you can take the class through the standard method for making a seven-color map on the torus. This construction involves using the square gluing pattern for the torus to construct a map with seven hexagons, each of which borders the other six (see ⌊5, p.63⌋);

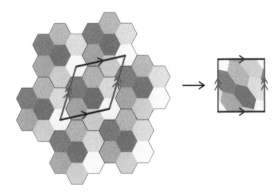

Figure 11. A seven-color map on the torus.

this gives a nice complement to the six-pentagon map on the projective plane. As shown in Figure 11, a doubly periodic hexagonal coloring of the plane can be cut and glued to make a seven-color torus map.

* In a graph theory course, you can elaborate on the theory of planar and non-planar graphs by identifying the antipodal points of an icosahedron. This gives the dual to the dodecahedron map, a configuration of six vertices with one edge between each pair of vertices. In other words, this gives a non-crossing complete graph K_6 on the projective plane (see Figure 12).

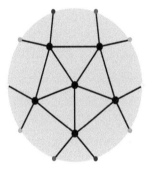

Figure 12. The complete graph K_6 on the projective plane (boundary points with matching colors are identified).

The reason that K_6 is projective-planar but not planar is that the Euler characteristic of the projective plane is different than that of the Euclidean plane. (For an explanation of the Euler characteristic of a surface, see Chapter 1.) In fact, Fortunatus's Purse gives a nice derivation of the Euler characteristic χ of the projective plane: since the purse has three faces, six edges, and four vertices,

$$\chi = 3 - 6 + 4 = 1,$$

as opposed to the Euclidean plane's Euler characteristic of 2. This calculation dovetails nicely with a proof using Euler's formula [2, p.106] that K_5 is a non-planar graph.

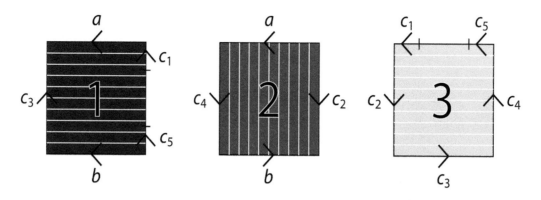

Figure 13. The stitch pattern for Fortunatus's Purse.

4 How to Make Fortunatus's Purse

The basic instructions for how to make Fortunatus's Purse are simply to follow the directions given by Mein Herr at the beginning of this chapter, except to sew along the entire first seam instead of just at the corners. However, some thought must go into what type of cloth squares to use and how to stitch them up. In particular, since every seam will be viewed from both sides, it is important that the seams be reversible.

4.1 Pattern Overview

Regardless of the type of cloth squares and method of stitching them, the following overall directions hold. Choose three squares of fabric in complementary colors that are easy to tell apart. Stitch one full edge of the first square to one full edge of the second square. Then take the opposite edges of each of these squares and stitch them with a half-twist to form a Möbius band of fabric. These are marked as stitches a and b in Figure 13.

Now you have reached the notorious impossible step, so you need to allow for a gap in the stitches around the third handkerchief. To make the model neater and more symmetric, begin stitching one quar-

ter of the way along one edge of the third handkerchief and one quarter of the way along one of the unstitched edges of the first handkerchief. When you reach the nearest corners of the first and third handkerchiefs, you will be able to stitch the next edge of the third handkerchief to the adjacent unstitched edge of the second handkerchief. Continuing around the third handkerchief, you will reach the remaining unstitched edges of the first and second handkerchiefs in turn, and then you will arrive at the other corners of the original edges of the first and third handkerchiefs. Stitch one quarter of the way into these edges, tie off, and your purse is complete! This sequence of stitches is marked as c_1 through c_5.

4.2 Fabric Squares

There are two possible sewing methods for Fortunatus's Purse: machine stitching and hand stitching. Each method has advantages and disadvantages, and each requires a particular type of fabric square.

With a sewing machine, you need to sew without overlapping the fabric squares for a completely invertible seam. This requires the use of a non-raveling fabric such as fleece or felt, and, naturally, the fabric itself should also be invertible. As it happens, fleece Fortunatus's purses make rather fetching hats, and a purse sewn

Figure 14. Fortunatus's Hat?

from 13″ squares of fabric fits the average adult head nicely (see Figure 14).

For hand sewing, it is best to use pre-hemmed squares of fabric, since these provide smooth, sturdy edges to stitch. One natural option for making the purse is to take the traditional approach and use handkerchiefs. I must confess that I have never tried this myself, as I suspect that even if it were easy to find handkerchiefs in a fixed size and contrasting colors, the fabric itself would be too thin to stand up to repeated handling. To be fully appreciated, Fortunatus's Purse should be turned every which way by everyone who picks it up. For this reason, I prefer to use cloth dinner napkins. Most home stores will carry a given brand and style of dinner napkin in a variety of tasteful colors. As mentioned above, another advantage of using dinner napkins is that most of them have visible hems, which allow the viewer to distinguish the two sides of each cloth panel. Napkins with a coarse weave are also useful for gauging the size of your stitches to keep them uniform.

Some napkins, like those in the two purses pictured in this chapter, have a pronounced grain parallel to one pair of edges. If this is the case, the most symmetrical arrangement is to have each napkin meet each other napkin so that their grains are perpendicular to each other, as diagrammed in Figure 13 and photographed in Figures 15 and 17.

The primary advantages of machine sewing are speed and flexibility. With handcut squares of fabric, you have more control over the size of the purse and a wider selection of colors and patterns. On the other hand, using a sewing machine gives you less control over the stitches themselves. Hand stitching produces a more polished and elegant seam, and there is something satisfying about using a more Victorian technique. Sewing the purse by hand takes much, much longer than sewing it by machine, however, and using napkins or other prefabricated squares introduces another complication. Even within a given brand, there will always be a slight variation in the lengths of the sides. By turning the squares around you can attempt to minimize the discrepancies, but in most cases you will have to do a little surreptitious gathering when you stitch.

4.3   Stitches

To make a machine-sewn purse, choose a neutral color of thread and use a wide zig-zag stitch, or any wide stitch that looks the same from the back as from the

Figure 15. Overcast stitch.

front. Feed the squares of fleece or felt into the machine side by side with no overlap between the two pieces. As the purse progresses, the task of ensuring that only the layers of fabric that you are currently stitching wind up under the presser foot grows more challenging, but as long as you keep the bulk of the purse bunched up behind the foot and rearrange the purse as you go, you will have no trouble completing all of the seams.

In a hand-sewn purse, there is no easy way to make the stitches inconspicuous. The natural solution is to use ornamental stitching, which suggests a thicker thread with a color that complements all three fabric colors. Embroidery floss works nicely, especially when paired with woven dinner napkins. There are two good stitches, each with its own advantages and disadvantages.

The first option is a narrow overcast stitch. While this stitch is performed asymmetrically, if you stay close to the edge of the fabric, you can pull the seams flat after they are sewn to distribute the thread evenly on both sides of the fabric. For each seam, pin the two fabric edges together, starting with the ends, the middles, and the quarter points. This will allow you to distribute any length discrepancy all along the seam.

The stitch itself is quite simple: on each pass, pull the needle front to back through both layers of fabric, let-

ting the thread wrap over the top to bring the needle over to the front again, as in Figure 16.

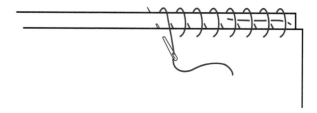

Figure 16. Making the overcast stitch.

Keep each stitch 1/8″ or closer to the edge and no more than 1/8″ wide. This is painstaking work, but it ensures an even seam that will look roughly the same from the underside. If you are working with a coarsely woven napkin, you can measure the stitches against the weave of the napkin itself, almost as though you were needle-pointing. If one of the layers is slightly longer than the other, even them out by periodically pulling the needle through at a slant to push a little more of the long side behind the needle, making sure that the layers are properly aligned by the time you reach the next pin.

Make the first few stitches in each row around the tail of the thread to cover it up as shown in the diagram. Similarly, at the end of each row, pull the needle back through your final stitches to secure the end of the thread before you cut it off. Although Figure 16 shows a single thread for clarity, I recommend using a double thread to produce a solid seam with fewer stitches.

Figure 17. Baseball stitch.

The second option is a baseball stitch. This is a fully reversible stitch that looks quite elegant, but it requires more skill than the overcast stitch and is not as amenable to subtle gathering. In the baseball stitch, the two pieces of fabric are laid flat with the edges abutting. (See Figure 17.) To make a stitch, pull the needle through one piece of fabric from the top to the bottom 1/8″ from the edge, then draw the needle back to the top between the fabric pieces. The next stitch is through the second piece of fabric 1/8″ from its edge, and the remaining stitches alternate from one side to the other, as in Figure 18.

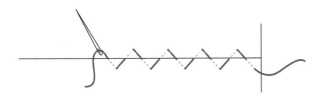

Figure 18. Making the baseball stitch.

Consecutive stitches are about 1/4″ apart in the direction of the seam. Unfortunately, pinning the two fabric edges together is ineffectual for the baseball stitch, and it is very easy to overcompensate if you try to gather a longer edge throughout the stitching process. Instead, stitch the first three quarters of the edge perfectly flush,

and then shorten the stitches on the short side, rechecking the rest of the edge every few stitches until the corners are properly aligned. Secure the tails of thread at the beginning and end of the seam by tucking them into the hem of the napkin.

A note of warning: either stitching method will take longer than you expect, so be sure to allow yourself plenty of time! After all, no one said getting unlimited wealth would be easy.

Bibliography

[1] Carroll, Lewis. *Sylvie and Bruno Concluded*. Dover Publications, New York, 1988.

[2] Trudeau, Richard J. *Introduction to Graph Theory*. Dover Publications, New York, 1994.

[3] Wilson, Robin. "Alice in Numberland: An Informal Dramatic Presentation in 8 Fits." *The College Mathematics Journal*, vol. 3, no. 5 (November 2002), pp. 354–377.

[4] Wilson, Robin. *Four Colors Suffice*. Princeton University Press, Princeton, NJ, 2003.

[5] Wilson, Robert A. *Graphs, Colourings, and the Four Color Theorem*. Oxford Univ. Press, Oxford, UK, 2004.

CHAPTER 8

(k)not cables, braids

SARAH-MARIE BELCASTRO
AMY F. SZCZEPAŃSKI
CAROLYN YACKEL

1 Overview

Knitted sweaters are frequently embellished with beautiful textures worked into the design. Cabling is a popular technique for creating such texture; knitted cables are patterns of raised stitches that cross over and under each other as in Figure 1.

Cable patterns are the hallmark of Irish fisherman knits, and stitches that cross and weave are also found in some traditional Estonian knitting [3] as well as in designs from Bavaria, Tyrol, Alsace, Norway, and Denmark. It makes sense that these patterns would be popular in cold climates, because cabling uses more yarn than flat (e.g., stockinette) knitting, so the resulting garments are thicker and warmer.

To knit a cable pattern, one physically moves sets of (usually two to four) contiguous stitches in front of or behind other stitches. The exchanging of strand groups, weaving sets in and out, will be familiar to anyone who has attempted to braid hair. Indeed, mathematicians use the term *braids* to describe cables. The two notions are quite similar. Both mathematical braids and typical cable patterns follow the progress of several vertically oriented strands as they weave back and forth and over and under each other. For most knitted cablework, the directions are given in terms of a chart that diagrams how the stitches should be worked in a vertically repeated pattern block or motif. Similarly, mathematicians use small geometric elements that are stacked vertically when describing a specific braid. This gives us a very natural way to use the language of mathematics to describe and analyze cables, as we will do in the next section.

In Section 2.1 we create a dictionary to translate between the language of cables and the language of braids; then we describe the mathematical structure of braids in Section 2.2. One of the big mathematical ideas represented here is that of braid *equivalence*, or how to tell when two different-looking knitted cables represent the same mathematical braid.

While most crafters loathe knots in their yarn and floss, mathematicians have a much more pleasant take on *knots*, which may be constructed from braids. The mathematics of knots is discussed more in Section 2.3. These knots (and links) can arise naturally in knitting. Although cables on sweaters are usually arranged to run up and down on both the body and the sleeves, as in Figure 2, nothing stops us from having a cable go

Figure 1. A few cable patterns.

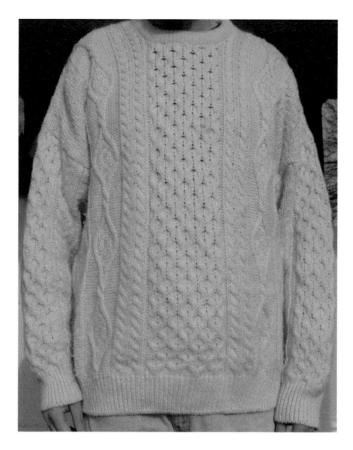

Figure 2. A fisherman's sweater from Ireland.

around the body and connecting the beginnings of the strands to their ends. This could be accomplished by choosing a cabled design and knitting it around the cuffs or around the bottom of a sweater. (See, for example, the chapter "Fringes" in [4].) This type of design is what mathematicians call a closed braid. These tangled strands correspond to a knot if the pattern can be completely traced by following one strand around. If following just one strand leaves some parts of the pattern untraced, the closed braid is a *link*, meaning it is more than one knot, possibly looped together. Whimsical linked and knotted constructions can be found in [5] in the chapter "Cords." Other knitters have found ways to include knotwork in their projects [6, 8].

In Section 3, we give several activities that may be used to investigate mathematical braids using knitted cables. The project in Section 4 uses a set of cables that look different but are all mathematically equivalent as braids. This is explained at the end of Section 2.2, so keep reading!

2 Mathematics

As mentioned above, braids and cables are essentially the same. Mathematicians formally define a braid to be a set of n strands that begin at some horizontal line and end at some other horizontal line, moving only in the same vertical direction, although they may move over and under each other laterally. One result is that any horizontal plane intersecting the braid intersects each strand exactly once. While mathematicians typically read braids from top to bottom to align with gravity, in this chapter we will read them from bottom to top in order to be consistent with knitting charts. In addition to drawn braids, mathematicians use compact notation. Though the notation varies according to the source, we present the standard notation, as given in [1], with the caveat that the braid first be rotated by π since we are reading bottom to top. Let the n strands of the braid be numbered $1, 2, \ldots, n - 1, n$ from right to left. If there are no twistings of the strands in the braid, it's not very interesting, so we will notate nothing. That is, we will write nothing down to describe this non-braid. A twist of strand i in front of strand $i+1$ is denoted σ_i, and a twist of strand $i + 1$ in front of strand i is denoted σ_i^{-1}. Note that these twists are actions, so they refer to whichever strands happen to be in the ith and $(i + 1)$st positions just before the twist occurs rather than the original ith and $(i + 1)$st strands. An example is given in Figure 3.

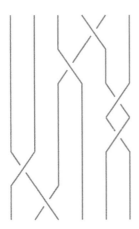

Figure 3. This braid is "read" as $\sigma_4\sigma_5^{-1}\sigma_1^{-1}\sigma_1^{-1}\sigma_3\sigma_2$.

We can *concatenate* two braids on the same number of strands by simply placing one above the other and gluing the ends of the strands of the first braid to the starts of the strands of the second braid. In fact, braids on n strands form a group B_n under this concatenation

operation. The σ_i generate B_n, but σ_i only commutes with σ_j when $|i - j| > 1$, because then none of the same strands are involved in the two twists. An element of B_n is known as a *braid word*, where the "letters" are the twists of the braid.

2.1 Braids versus Cables

The equivalence between knitted cables and mathematical braids is not quite exact. For example, many different cables may represent the same braid, since each cable represents a different braid drawing. However, there are many cables that don't follow the usual mathematical rules for braids. Here we attempt to expand the mathematics-knitting dictionary to translate between cable patterns and braids.

⋆ In a mathematical braid, all the strands have the same width. In knitting, there are cables with different stitch-widths, and thus different strand-widths. For a cable with all strands of the same width, we merely specify how many stitches there are per strand. This methodology also works for cables with strands of varying widths, as long as we choose the number of stitches per strand to be the greatest common divisor of the cable's stitch widths. Likewise, to complete the mathematics-knitting translation for these braids, we need to specify the height of a cable, that is, how many rows should be knitted before another strand-twist happens.

⋆ Sometimes a cable will have two independent strand-twists happen at the same time. Mathematically, it does not matter which one we write first in the braid word because of the commutativity noted above. However, the braid word may not indicate the cable pattern we began with unless we introduce new notation. To denote the fact that these independent strand-twists happen at the same time in a cable pattern, we set them off in parentheses. For example, in

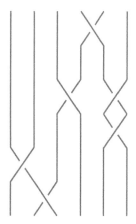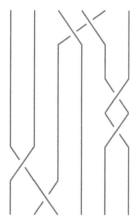

Figure 4. Slightly altered versions of Figure 3 with simultaneous strand-switches, corresponding to the braid words $\sigma_4\sigma_5^{-1}\sigma_1^{-1}(\sigma_1^{-1}\sigma_3)()\sigma_2$ (left) and $\sigma_4\sigma_5^{-1}\sigma_1^{-1}\sigma_1^{-1}()(\sigma_3\sigma_2)$ (right).

Figure 4 (left) we have a cable pattern corresponding to the braid word $\sigma_4\sigma_5^{-1}\sigma_1^{-1}(\sigma_1^{-1}\sigma_3)()\sigma_2$. The () in the braid word signifies that there is a repeat of one cable-height in which no cabling takes place.

★ There are many different knitted cables that are all equivalent to the trivial braid; two are shown in Figure 5. Some of these have no crossings—the strands move back and forth across the knitted fabric, but never cross each other. How to include such cables in the knitting-mathematics translation dictionary is an open question.

2.2   The Braid Group

The braid group B_n surjects naturally to the symmetric group S_n on n letters. The reader should note that while σ_i is an involution in S_n, it is not a involution in B_n, as we know from the experience of repeatedly twisting two strands. Indeed, σ_i does not have finite order in B_n.

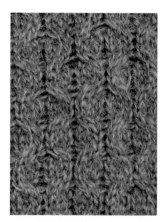

Figure 5. Both of these cables represent the trivial braid.

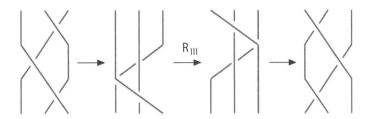

Figure 6. The braid word $\sigma_i\sigma_{i+1}\sigma_i$ is equivalent to the braid word $\sigma_{i+1}\sigma_i\sigma_{i+1}$.

Yet, B_n is not simply the free group on the σ_i. As we noted above, $\sigma_i\sigma_j = \sigma_j\sigma_i$ for $|i - j| > 1$. Furthermore, we have analogues of the Reidemeister moves, as defined for knots. The cancellation of $\sigma_i\sigma_i^{-1}$ from a braid word produces a Type II Reidemeister move in a drawing of the braid. There is also a Type III Reidemeister move that corresponds to the algebraic substitution of $\sigma_i\sigma_{i+1}\sigma_i$ for $\sigma_{i+1}\sigma_i\sigma_{i+1}$, as shown in Figure 6.

Using both of these algebraic substitutions gives a different Type III Reidemeister move, as in Figure 7. Note that no Type I Reidemeister move can be performed on a braid as this would require a braid strand to change direction.

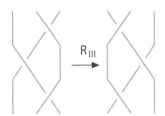

Figure 7. The braid-word $\sigma_i^{-1}\sigma_{i+1}\sigma_i$ is equivalent to the braid-word $\sigma_{i+1}\sigma_i\sigma_{i+1}^{-1}$.

It turns out that these three braid-word relations, $\sigma_i\sigma_i^{-1} = e$, $\sigma_i\sigma_{i+1}\sigma_i = \sigma_{i+1}\sigma_i\sigma_{i+1}$, and $\sigma_i\sigma_j = \sigma_j\sigma_i$ when $|j - i| > 1$, completely characterize braid-word equivalence [1, Section 5.4].

The braid-word relations are used in the project pattern: it consists of a series of panels, each containing a cable corresponding to a braid. The braids are all equivalent in B_n. In fact, reading the panels left to right across

the design results in a series of seven moves, as can be seen in Figures 17–20. Between panels 1 and 2 we apply a Type II Reidemeister move. Between panels 2 and 3 we apply the first sort of Type III Reidemeister move described above, and as shown in Figure 6. Between panels 3 and 4 and between panels 7 and 8, we use $\sigma_i\sigma_j = (\sigma_j\sigma_i)$ for $|j - i| > 1$. The move connecting panels 4 and 5 is used between panels 7 and 8 as well. It amounts to nothing more than collapsing the figure where no switching of strands occurs. This leaves only the moves between panels 5 and 6 and between panels 6 and 7. Both of these moves correspond to the Type III Reidemeister move presented in Figure 7. The original panel braid was $\sigma_3^{-1}\sigma_3\sigma_2\sigma_2\sigma_3\sigma_2\sigma_1\sigma_2\sigma_3^{-1}\sigma_2\sigma_1^{-1}$, and the final panel braid is $\sigma_2(\sigma_1^{-1}\sigma_3)\sigma_2(\sigma_1\sigma_3)\sigma_2\sigma_2\sigma_1^{-1}$. The careful reader will note that the first panel could not have been collapsed vertically without using the Reidemeister moves. Therefore, it had to have at least eleven units available for cabling switches, yet the final panel needed only seven vertical units.

2.3 Braids and Knots

It's easy to convert a braid into a knot or link. Just identify the end of the ith strand on the top with the end of the ith strand on the bottom, as in Figure 8. This is called a *closed* braid.

It's less clear how to turn a generic knot or link into a braid, but it can always be done (see [2] for details).

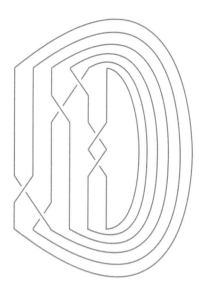

Figure 8. The braid from Figure 3 turned into a knot.

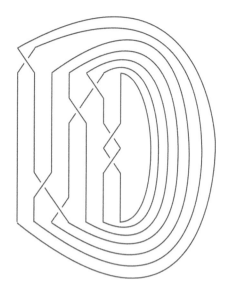

Figure 9. What happens when the closed-braid word from Figure 8 is multiplied by σ_6.

Notice that depending on where we choose to "start" and "stop" a closed braid, we will obtain different braid words. In fact, any two closed-braid words that differ only by a cyclic shift correspond to the same closed braid. Thus conjugating a braid word by σ_k (that is, premultiplying by σ_k^{-1} and postmultiplying by σ_k) is essentially the same as multiplying the corresponding closed-braid word by the identity. In terms of a braid drawing, such a conjugation places a σ_k^{-1} strand-twist below the braid and a σ_k strand-twist above the braid.

Another algebraic operation that has no effect on a closed braid is multiplying a closed-braid word on n strands by σ_n (or equivalently by σ_n^{-1}). This performs a Type I Reidemeister move (finally!—you knew there had to be one in here somewhere) on the nth strand. This can be seen in Figure 9; alternatively, we may view this operation as adding a $(n + 1)$st strand and connecting it to the closed braid so that it does not form a non-equivalent link. This operation is known as *stabilization*, perhaps because it stabilizes the equivalence class of our original braid within the braid group.

Fascinatingly, conjugation, stabilization, and the three braid-word relations suffice to distinguish inequivalent closed braids [7].

2.4 Braids and Symmetry

The standard way of plaiting a three-strand braid (such as commonly used when braiding hair) is to repeat the element $\sigma_1\sigma_2^{-1}$. Generalizations of this braid to n strands are possible, and one n-strand braid uses weaving. Take the first strand and weave it over the second, under the third, over the fourth, and so on, until all the other strands have been crossed; repeat this with the strand that is now the first strand. This can be denoted as taking powers of the element $\sigma_1\sigma_2^{-1}\sigma_3\sigma_4^{-1}\ldots$. For example, plaiting a six-strand braid would take powers of $\sigma_1\sigma_2^{-1}\sigma_3\sigma_4^{-1}\sigma_5$, and a seven-strand braid would take powers of $\sigma_1\sigma_2^{-1}\sigma_3\sigma_4^{-1}\sigma_5\sigma_6^{-1}$. (This is how Amy conceptualizes braiding hair.) However, this is not the way that braids are formed in knitted cablework, though the only difference is in the ordering of the elements.

These braids usually feature the product of all of the odd-index elements followed by the product of the inverses of all of the even-index elements. In this method a six-strand braid would be $\sigma_1\sigma_3\sigma_5\sigma_2^{-1}\sigma_4^{-1}$ and a seven-strand braid would be $\sigma_1\sigma_3\sigma_5\sigma_2^{-1}\sigma_4^{-1}\sigma_6^{-1}$. (In contrast, this is how sarah-marie conceptualizes braiding hair.)

When $n = 3$, these two methods are identical, but for $n > 3$ they are not.

3   Teaching Ideas

The braid group can provide useful examples in any class that explores algebraic properties such as the commutative and associative laws. Working with the braid group appeals to students with a variety of learning styles. Students with a visual learning style may wish to make diagrams of the braids, while those with a more hands-on style can work with strands of wire or string.

For informal experiments, it's helpful to have a supply of several colors of yarn or twine. Using different colors for each strand makes it easier to keep track of the permutation generated by the braid. In that case, however, be sure the students understand that for a fixed i, σ_i may act on different colors at different stages of the braid. A configuration made from fiber may be preserved by taping it to an index card. To create more stable models of a braid diagram, use colored fine-gauge electrical wire (such as the wires found inside telephone or networking cables). The ends of the wire can be wrapped around a narrow dowel (or a pencil) to keep the braid from unraveling, or the ends of the braid can be attached to plastic needlepoint canvas.

To have a common language for describing braids and cables, students should first be introduced to notation for the generators of the braid group and the conventions that apply. Advanced students who are able to read articles in the scientific literature should be warned that authors vary in the way they draw and read the braid diagrams (top to bottom, bottom to top, or left to right) and whether σ_i denotes strand i crossing *over* strand $i + 1$ or *under* strand $i + 1$. Once students can consistently write down the algebraic notation for a braid diagram and translate a braid diagram to its algebraic notation, they can begin to explore the properties of this group.

3.1   Working with Basic Properties of Braids and Knots

These questions can be suitably rephrased to be grade-level appropriate for secondary through graduate students.

★ Draw three different braids that all have the same number of strands. Concatenate any two of the braids. Now concatenate them in the other order. Do you get the same braid as a result? Does the order of concatenation behave the same way no matter which two braids you pick?

★ Choose a braid you have previously drawn. Can you create a second braid that "cancels out" your chosen braid when you concatenate the two braids?

★ Use Reidemeister moves to show that the braid relations given in Section 2.2 hold.

★ Can you find an element of B_n that, when converted to a closed braid, is a knot?

★ Working the other way, given a knot or link (such as might be found in Celtic knotwork), can you find a braid that represents the knot?

★ Can you find elements of B_n that are links with two components? With any number of components up to n?

★ How does B_n naturally sit inside of B_{n+1}? Are there more interesting homomorphisms of braid groups? Can you find a homomorphism from the braid group B_n onto the symmetric group S_n?

★ What does a normal subgroup of a braid group act like in terms of the strands of the cables?

The project given in Section 4 has been designed to exemplify a range of braid properties. Begin by having students create braid drawings from the cable charts. Then, they can verify that these braid drawings

Figure 10. Three mathematically nontrivial (and different) braids. Can you express these using notation in B_n?

show topologically equivalent braids. Students can be asked to identify Reidemeister moves in the braid drawings. For an algebraic perspective, ask students to write down the braid words associated to the braid drawings and identify the braid-word relations that correspond to moving adjacent drawings of panel braids.

3.2   Braids in the Wild

Since cable patterns commonly decorate knitwear, ask students to locate scarves, sweaters, or other items featuring cable patterns and investigate the corresponding braids. If the students have trouble finding garments featuring cable patterns, they can work from photographs that clearly show the cabling (such as in a knitting book about Aran sweaters); a few examples are given in Figure 10. Ask the students to use mathematical notation to record which elements of the braid group are found on the items.

Once the class has collected an assortment of cable patterns, the students can begin to look for features that are common (or uncommon) among the patterns. How many strands are in each cable? Is there an even number of strands or an odd number of strands? How often is each of the generators σ_i represented in each cable? If a sweater contains more than one type of cable, is there a relationship among the different cables in the sweater?

How often do the cables represent the trivial braid?

Once the students have determined what is "usual" and "unusual" for braid patterns that appear on garments, they can be asked to design their own patterns. Here are some sample exercises.

* Design a cable pattern with the same braid word as a cable you've seen in a garment, but that has a different appearance when realized as knitting.

* Design a cable pattern with a braid word you haven't seen before, but that has features in common with the patterns you have seen in garments.

* Design a cable pattern whose mathematical description is outside the norm of what you've seen on garments.

3.3   Wildly Braiding

These questions relate to the symmetry of the two methods of braiding introduced in Section 2.4.

* Discuss why the action $\sigma_1\sigma_3\sigma_5\sigma_2^{-1}\sigma_4^{-1}\sigma_6^{-1}$ is not symmetric when the braid has an even number of strands, but *is* symmetric when the braid has an odd number of strands.

* Investigate to what extent this symmetry (or lack of symmetry) extends to a diagram of a braid.

* Which of these braids, $\sigma_1\sigma_3\sigma_5\sigma_2^{-1}\sigma_4^{-1}\sigma_6^{-1}$ and $\sigma_1\sigma_3\sigma_5\sigma_2^{-1}\sigma_4^{-1}$ (and their generalizations), can be drawn in such a way so that the diagram has reflectional symmetry? Do they have rotational symmetry in any sense?

* Do the two methods of constructing braids have comparable results? For a three-strand braid, they are exactly the same, but what happens when $n > 3$?

* Consider the two methods of braid construction from Section 2.4 on a fixed number of strands. Now consider forming the result of each into a closed braid. Are the two closed braids equivalent? If this seems difficult, focus on the cases of four- and five-strand braids.

4 How to Make a Pillow of Braid Equivalence

These instructions are for making a pillow cover as shown in Figure 8, but the motif given in the cable charts would work just as well for embellishing other objects. It could be made into wall art (see Figure 14) or part of an afghan; divided into two halves (cable panels 1–4 and 5–8), the motif would work as the front and back of a bag or sweater body or two ends of a horizontally knitted scarf.

Materials

* One size 9, 40″ circular needle on which we will knit back and forth.

* Exactly two 100-gram skeins of Blue Sky Organic Cotton (150 yards each) in your favorite color. We used sage.

* One 12″ × 16″ pillow.

or

* One size 7 circular needle on which we will knit back and forth.

* Exactly two 50-gram skeins of Valley Goshen (92 yards each) in your favorite color. We used sage.

* One 9″ × 12″ pillow or a pile of fiberfill.

Figure 11. The pillow of braid equivalence, front (left) and back (right).

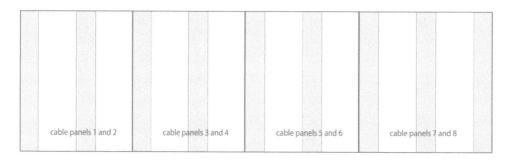

Figure 12. How the cable panels fit together.

Gauge

* 4.25 stitches per inch by 5 rows per inch (Blue Sky Organic Cotton, abbreviated BSOC)

or

* 4.5 stitches per inch by 6 rows per inch (Valley Goshen, abbreviated VG).

Instructions

Cast on 150 stitches. Work 9 rows in garter stitch (that is, knit 9 rows) if knitting with BSOC, or work 13 rows in garter stitch (that is, knit 13 rows) if knitting with VG. Beginning with row 10 (for BSOC) or row 14 (for VG), follow the cabling charts provided in Figures 17–20 on pages 130–133. The symbols in the cabling charts are given in Figure 16.

Notice that row 10 (or row 14) is worked on the wrong side, so that the pattern for this row is *P1 K1 P1 K1 P1 K1 P12* P1 K1 P1 K1 P1 K1; this allows the cabling to happen only on right-side (primarily knit) rows. This means that you must read the first row of the chart in the direction opposite what you normally do. The cable charts in Figures 17–20 direct the work for 44 rows (rows 10 through 53 for BSOC; rows 14 through 57 for VG). Figure 12 shows how the four charts fit together.

Figure 13. A variant pillow with purl columns, front (left) and back (right). This pillow was constructed using hemming instead of lapping.

Figure 14. A wall hanging with internal purl columns and external garter columns.

Two attractive variants on the cable charts are to replace the seed-stitch columns with purl columns or garter columns. (If you substitute purl columns, you may wish to do seed stitch or garter stitch on the first and last three stitches of each row in order to avoid edge curling.) Figure 13 shows a pillow with purl columns, and Figure 14 shows a wall hanging with purl stitches used for the inner columns and garter stitch used for the two outer columns.

Work 8 more rows in garter stitch if using Blue Sky Organic Cotton, or 12 more rows in garter stitch if using Valley Goshen. These are rows 54 through 61 for BSOC or rows 58 through 69 for VG. Bind off.

Now fold the knitted piece as shown in Figure 15. Note that this folding is intentionally asymmetrical; it can be done symmetrically, but the lapped area will be more difficult to sew. Match the short ends together with right sides facing in, then lap the last six seed stitches beneath the first six seed stitches. Using the left-hand edge of the work as a guide, flatten the knitting into a rectangle. (Alternatively, hem the first and last three stitches, align the hemmed edges, and flatten the knitting into a rectangle.) The right-hand fold edge will be between the fourth cable panel and the fifth seed column. Whipstitch or mattress-stitch across the top and across the bottom. Turn right side out. Insert the pillow. Now, take a nap.

- ☐ knit on right side, purl on wrong side (stockinette)
- ⊙ purl on right side, knit on wrong side (reverse stockinette)
- cable 3 stitches to the right
- cable 3 stitches to the left

Figure 16. The key to the cable panels.

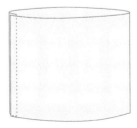

Figure 15. Lapping the work before sewing.

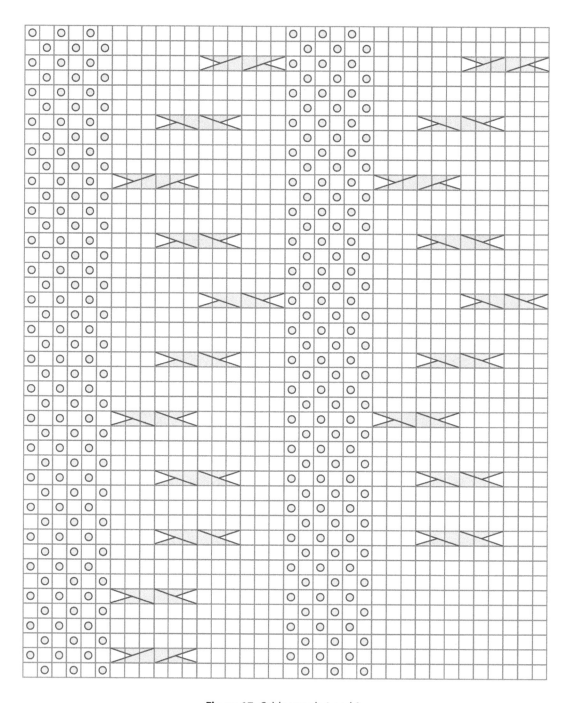

Figure 17. Cable panels 1 and 2.

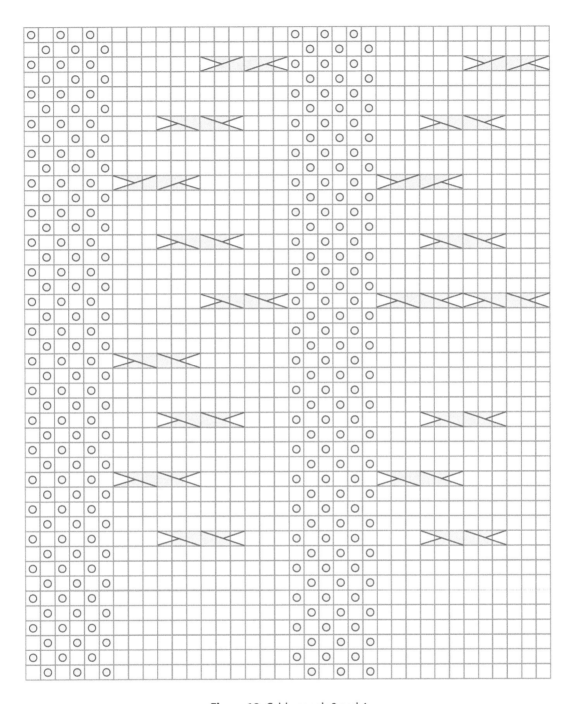

Figure 18. Cable panels 3 and 4.

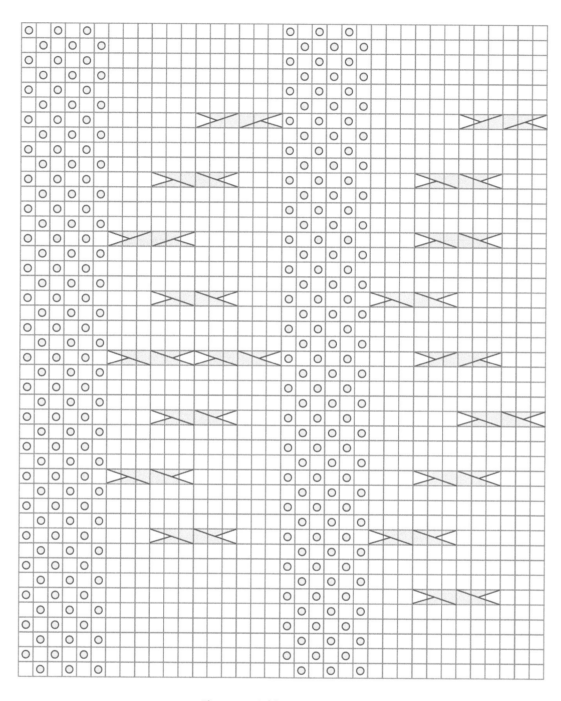

Figure 19. Cable panels 5 and 6.

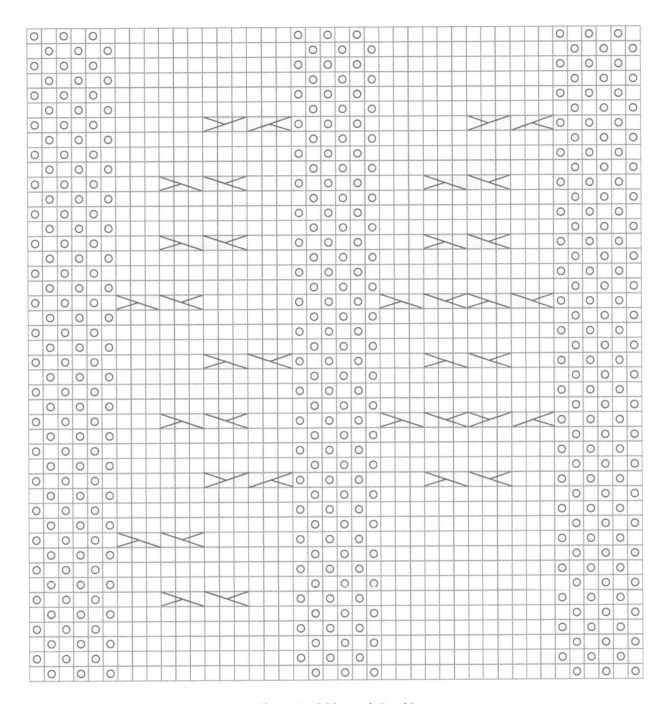

Figure 20. Cable panels 7 and 8.

Bibliography

[1] Adams, Colin. *The Knot Book.* W. H. Freeman, New York, 1994.

[2] Alexander, J. W. "A Lemma on Systems of Knotted Curves." *Proceedings of the National Academy of Sciences USA*, vol. 9, 1923, pp. 93–95.

[3] Bush, Nancy. *Folk Socks: The History & Techniques of Handknitted Footwear.* Interweave Press, Loveland, CO, 1994.

[4] Epstein, Nicky. *Knitting on the Edge: Ribs, Ruffles, Lace, Fringes, Flora, Points and Picots.* Sixth&Spring Books, New York, 2004.

[5] Epstein, Nicky. *Knitting over the Edge: Unique Ribs, Cords, Appliqués, Colors, Nouveau.* Sixth&Spring Books, New York, 2005.

[6] Lavold, Elsebeth. *Viking Patterns for Knitting: Inspiration and Projects for Today's Knitter.* Trafalgar Square Publishing, North Pomfret, VT, 2000.

[7] Markov, A. A. "Über die freie Äquivalenz der geschlossenen Zöpfe." *Recueil Mathematique de Moscou*, vol. 1, 1935, pp. 73–78.

[8] Wilson, Jenna. *The Girl from Auntie*, "Free Cable Patterns." http://www.girlfromauntie.com/patterns/celtic/, 2006.

CHAPTER 9

the graph theory of blackwork embroidery

JOSHUA HOLDEN

1 Overview

1.1 A Brief History of Blackwork

Blackwork is an embroidery technique generally associated with Tudor England. In that period and place, it was frequently done with black thread on light colored linen, hence the name. Other distinguishing features include a preponderance of straight lines and geometric shapes, which make it eminently suitable for mathematical treatment.

The origins of blackwork, as with many crafts, can be best described with the cliché "lost in antiquity." Legend has it that the technique was brought to England from Spain around 1501 with the court of Catherine of Aragon. Catherine was the intended bride of Prince Arthur, heir to the throne of England. When Arthur died a year later, Catherine was betrothed to the new heir, Arthur's younger brother Henry. Henry would later become Henry VIII, and Catherine would become the first of Henry's famous six wives.

This story, while romantic, is certainly false. It is true that techniques that could be called blackwork can be found in Spanish embroidery before 1501. However, they also appear in England and many other countries before Catherine's time. Despite that, the association with Catherine led to the use of the term "Spanish work" as another term for blackwork, and certainly gave a boost to the popularity of blackwork in England during this period.

We also associate blackwork with the early 1500s because of the paintings of Hans Holbein the Younger, court painter to Henry VIII. Holbein's paintings are so realistic that the stitching on the sleeves and collars of his subjects is clearly identifiable as blackwork and, in fact, can be reproduced by modern stitchers. Holbein's portrait of Jane Seymour ([6], also page 8 of [2]) is a good example of such a painting, and a chart of the pattern on Queen Jane's cuff can be found on page 105 of [2].

Figure 1. A modern blackwork sampler, front (left) and back (right).

Blackwork dropped out of favor in England during the seventeenth century, and other stitching techniques took its place [1, 8]. In the late twentieth and early twenty-first centuries blackwork has enjoyed a revival among embroiderers, historical reenactors, interior decorators, and others who enjoy its combination of art, technical challenge, and intellectual creativity. Modern stitchers can choose from a large number of patterns, both those reproduced from period examples (either from paintings or artifacts) and newly created modern designs. Modern patterns may follow the traditional colorways of black, or sometimes red, thread on a light-colored background. Or, they may use other colors including light thread on a black or dark background. Many examples of each may be found in [1]. For more on the history and practice of blackwork, see [8] and the references in its bibliography.

1.2   How to Stitch Blackwork

In addition to the high-contrast color schemes and straight-line geometric patterns, there are specific methods of stitching that are characteristic of blackwork. Firstly, the linen or other cloth used (especially by modern stitchers) is selected for a regular grid of threads and holes, as can be seen in Figure 2. Unlike more free-form methods of sewing and embroidery, here the thread will pass only through these holes as it travels across the fabric.

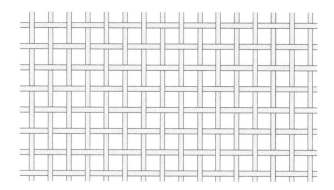

Figure 2. A typical piece of fabric used in blackwork.

Although any method of stitching that produces an unbroken straight line can be used for blackwork, the most traditional is the "double running stitch," also called "Holbein stitch" because of its association with Holbein's paintings. In double running stitch, the thread is brought up and down through the fabric while traveling in one direction ("running stitch"), as shown in Figure 3. (The numbers indicate in which order the thread should pass through the holes.)

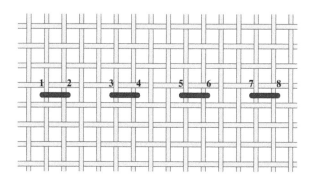

Figure 3. Running stitch.

This produces a broken line on the front of the fabric and a complementary broken line on the back. Then the direction is reversed and the line is retraced in the other direction ("double running stitch"), producing identical continuous lines on the front and back of the fabric, as in Figure 4.

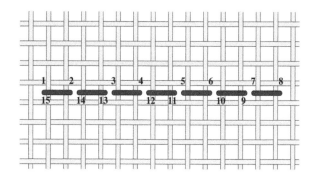

Figure 4. Double running stitch.

Branching paths can also be made by taking "side trips," as shown in Figure 5.

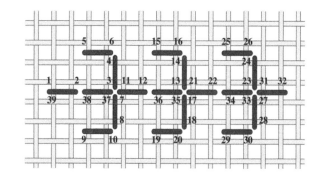

Figure 5. Double running stitch with "side trips."

Figure 6. A modern blackwork runner, front (top) and back (bottom).

If the path of the stitching is exactly reversed, then the finished stitching will be "reversible," having the same appearance from either side of the fabric. (Technically the pattern will be a mirror image, of course, but we will ignore this for the moment.) See Figure 6 for an example. This is one of the reasons blackwork was commonly used for sleeve cuffs and collars in the Elizabethan period. (Cuffs and collars in this period were generally ruffly, and both sides could be seen during normal use. The portrait of Jane Seymour [6], mentioned earlier, shows this well.)

It is also possible to use running stitch to produce pieces that have a different pattern on the front than on the back, or the same pattern shifted by some number of threads. Or, if the back of the piece will never be seen, it may not be considered important to preserve any pat-

tern at all on the back side. In this chapter, however, we will be interested primarily in "reversible" patterns stitched using double running stitch. For more on various stitches used in blackwork, [2].

1.3   Some Questions and Answers

Double running stitch creates as much of the pattern as possible with one continuous length of thread. Therefore, one question many beginning blackwork embroiderers ask themselves is: How do I know when to start and stop? The answer usually given by experienced stitchers is: Start anywhere, and stop when you run out of thread or finish the pattern. For a stitcher with a mathematical mind, however, this prompts more questions, such as: Can we prove that it doesn't matter where

Figure 7. A blackwork decorative box top—try to see how to stitch each connected piece using the accompanying algorithm.

we start? What if the thread is infinitely long—can we always finish the pattern without having to break the thread?

The answer to both of these questions is yes, as Section 2 will show in a precise mathematical fashion. To preview the main result, we will see that every connected blackwork pattern can be stitched reversibly with one thread. One method of doing so is given here.

1. Pick any point to start your thread, and come up at that point.

2. Stitch outward along any path until you come to a dead end or a point you have been through before.

   (a) If you have come to a dead end, turn around and stitch "back" to the next place you can branch off on a new path. Then go to Step 2.

   (b) If you have come to a point you have been through before, and you are stitching "outward," turn around and stitch "back" to the next place you can branch off on a new path. (This happens when you have gone around a closed loop in the pattern.) Then go to Step 2.

   (c) If you have come to a point you've been through, you are stitching "back," and you can't branch off on a new path, you should be able to keep stitching "back." If not, you are done with the pattern!

Try these steps for yourself on parts of the patterns in Figure 7. An example of a stitching path is given in Section 4 should you wish to look ahead.

2   Mathematics

2.1   Eulerian Graphs

The appearance of a blackwork pattern immediately brings to mind the mathematical concept of a graph. Unless noted otherwise, the definitions are more or less standard, but we present them as reminders and to fix the notation. (See, for example, [3, Chapter 11] or [4] for more information.)

A (loop-free) graph is a set of vertices, V, and a set of edges, E, where each edge is an *unordered* pair $\{x, y\}$ of distinct vertices. (This may sometimes be referred to as an "undirected" graph.) A (loop-free) directed graph, or *digraph*, is a set of vertices, V, and a set of edges, E, where each edge is an *ordered* pair (x, y) of distinct vertices. In a digraph, the order of vertices in each edge is often thought of as indicating a "direction" of travel, as in Figure 8.

Figure 8. A digraph.

A graph may be associated with any digraph by simply forgetting about the ordering of the pairs. Of course, such a graph may come from any of several different digraphs. The standard way of associating a digraph to a graph is by including in the digraph *both* possible orders (or directions) of each edge, as indicated in Figure 9.

We will think of our blackwork pattern as a digraph, where the direction of an edge is the direction of stitching. Note that alternate edges will end up on opposite sides of the fabric, as was shown in Figures 3 and 4. Thus every two adjacent vertices are connected by two

edges, one in each direction. Clearly a reversible blackwork pattern is a digraph that comes from a graph— each edge is traveled in one direction on the front of the fabric and the other direction on the back. (In fact, the digraph in Figure 9 is the same as the pattern in Figure 5.)

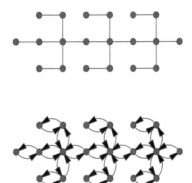

Figure 9. A graph (top) and its associated digraph (bottom).

Of course, this implies that the edges are traversed in some sort of order in the graph, as they are when one stitches. A *walk* on a graph is a finite alternating sequence of vertices and edges $x_0, e_1, x_1, \ldots, e_n, x_n$, where each $e_i = \{x_{i-1}, x_i\}$. A graph is *connected* if there is a walk between any two vertices. A *directed walk* on a digraph is a finite alternating sequence of vertices and edges $x_0, e_1, x_1, \ldots, e_n, x_n$, where each $e_i = (x_{i-1}, x_i)$. A digraph is *strongly connected* if there is a directed walk between any two vertices.

We can then make the following observations:

Observation 1 If a digraph is strongly connected then the associated graph is connected.

Observation 2 If a graph is connected then the associated digraph is strongly connected.

If we can stitch a reversible blackwork pattern with one thread then the pattern, regarded as a digraph, must be strongly connected. Our interest will be to see if every pattern that is a strongly connected digraph can be stitched reversibly with one thread.

Figure 10. A fancy blackwork box top.

It is usually considered aesthetically undesirable (i.e., ugly) to have more than one thread passing between the same two holes on the same side of the fabric. So we need to keep track of whether edges are repeated in our directed walks. A *directed trail* is a directed walk where no edge is repeated (although vertices may be). A *directed circuit* is a directed trail where $x_0 = x_n$. The *length* of a directed walk, trail, or circuit is the number of (not necessarily distinct) edges in it. If the length is zero, then the walk, trail, or circuit is called *trivial*. An *Eulerian circuit* of a digraph is a directed circuit that includes all of the edges of the digraph. A digraph is *Eulerian* if it has an Eulerian circuit. (In Figure 10, can you determine which components have Eulerian circuits?)

If we can stitch a blackwork pattern reversibly (with one thread) then the digraph must be Eulerian. (Of course, being Eulerian implies being strongly connected.) Now, it is well known which digraphs are Eulerian. For each vertex v of a digraph, the *outdegree* of v is the number of edges with v as the starting point; the *indegree* of v is the number of edges with v as the ending point.

Theorem 1 *A directed graph is Eulerian if and only if it is strongly connected and the indegree of each vertex is equal to the outdegree.*

This is the digraph version of Euler and Hierholzer's solution to the Seven Bridges of Königsberg. A proof of the undirected graph version can be found in [3, Section 11.3]; a somewhat different proof is given in [5, Chapter 7]. Adapting these to the directed graph version is not very difficult and is an interesting exercise for those with a little experience with graph theory.

In our situation we can apply the following observation:

Observation 3 Each vertex of a digraph associated to a graph has indegree equal to outdegree.

We then get a nice corollary to the theorem:

Corollary 2 *The digraph associated to a graph is Eulerian if and only if the original graph is connected.*

You may recall that we showed that a necessary condition for being able to stitch a blackwork pattern reversibly is that the digraph must be Eulerian. Unfortunately, we now see that this is not much of a condition, since every connected blackwork pattern is Eulerian. But this is not the end of the story!

2.2   Holbeinian Graphs

We have determined how to make sure that each edge in the graph is traversed exactly twice, in two different directions. However, we must still make sure that the two different directions of travel lie on opposite sides of the fabric. For that we need some more definitions. A digraph is *symmetric* if for every edge (x, y) in the digraph, the edge (y, x) is also in the digraph. Equivalently, a digraph is symmetric if it is the digraph associated to some graph. If $x_0, e_1, x_1, \dots, e_n, x_n$ is a directed trail on a digraph, we say that the *parity* of each edge e_i is the parity of i.

The next two definitions do not seem to be in the literature.

Definition 1 A *Holbeinian circuit* of a symmetric digraph is an Eulerian circuit where for all x, y the two edges (x, y) and (y, x) have opposite parities.

Definition 2 A symmetric digraph is *Holbeinian* if it has a Holbeinian circuit.

Note the analogy with the Eulerian case.

The parity of an edge corresponds to whether the thread is on the front side or the back side of the fabric, since double running stitches alternate between lying on the front and lying on the back. A blackwork pattern can be stitched reversibly with one thread if and only if the corresponding digraph is Holbeinian. (Surprise!) So, can we categorize Holbeinian digraphs?

Theorem 2 *A symmetric digraph is Holbeinian if and only if it is strongly connected.*

Corollary 3 *A symmetric digraph is Holbeinian if and only if the associated graph is connected.*

So it turns out that every connected blackwork pattern can be stitched reversibly with one thread, a fact that will come as no surprise whatsoever to any stitcher who has tried. Nevertheless, the proof of the theorem is interesting and perhaps instructive. It is closely based on the same theorem for the Eulerian case (see, for example, [3, Theorem 11.3]) with the addition of the parity condition. (An earlier, more algorithmic proof of this theorem was proposed by Lana Holden.)

Proof of Theorem 2: Let G be the symmetric digraph. If G has a Holbeinian circuit, then it must pass through every vertex. Thus for any two vertices, there is a directed walk from one to the other, so G is strongly connected.

Now assume that G is strongly connected, and let G^u be the associated graph. Let V be the vertex set of G^u

and E the edge set. If E has only one edge $\{x, y\}$, then $x, (x, y), y, (y, x), x$ is a Holbeinian circuit of G.

We will proceed by induction, supposing that the result is true if E has fewer than n edges. Consider a G^u in which E has n edges. Let $\{x, y\}$ be any edge of E. Let H be the digraph obtained from G by removing the edges (x, y) and (y, x). Now, either H is still strongly connected or it is divided into two strongly connected components, one of which contains x and the other y.

If H is still strongly connected, it has a Holbeinian circuit. (The circuit may be trivial if H consists of a single isolated vertex.) A Holbeinian circuit of G is constructed by following the Holbeinian circuit of H until the vertex x is reached, then following $x, (x, y), y, (y, x), x$, then finishing the Holbeinian circuit of H. The detour of length 2 does not affect the parity of any of the edges in the circuit of H.

If H is no longer strongly connected, each component has a Holbeinian circuit. Let the components be H_1 and H_2, with associated graphs H_1^u and H_2^u and associated undirected edge sets E_1 and E_2. A Holbeinian circuit of G is constructed by following the Holbeinian circuit of H_1 until the vertex x is reached, then following $x, (x, y), y$, then following the Holbeinian circuit of H_2 back to y, then following $y, (y, x), x$, and finally finishing the Holbeinian circuit of H_1. The detour of length $2 + 2|E_2|$ does not affect the parity of any of the edges in the Holbeinian circuit of H_1.

Note that passing from G to the components of H could be implemented in an algorithm as a recursion. (We will come back to this idea in Sections 3 and 4.) However, people don't naturally stitch recursively—they stitch iteratively, so we would like an iterative algorithm. In addition, this proof is "global"—it requires one to look at the whole graph and plot out a course. We are lazy and impatient and would prefer a "local" algorithm. Luckily, a French mathematician named Trémaux devised such an algorithm for the Eulerian case in the nineteenth century (see [7, Chapter 3]). Trémaux wanted to traverse a maze or a labyrinth. The goal was to visit every vertex. As a by-product, every edge is traversed twice. All that is needed is to add the parity condition.

Suppose that we have a strongly connected symmetric digraph.

1. Start at an arbitrary vertex x_0.

2. Proceed along any edge.

3. At each later step, suppose we have just traversed an edge (x, y) and arrived at a vertex y.

    (a) If we have not visited y before,

        i. if there is an edge leaving y other than (y, x), proceed along any such edge.

        ii. if there are no such edges, turn back along (y, x).

    (b) If we have visited y before,

        i. if we have not traversed (y, x), turn back along (y, x).

        ii. if we have traversed (y, x),

            A. if there is an edge (y, z) for which neither (y, z) nor (z, y) has been traversed, proceed along any such edge.

            B. if every edge (z, y) has been traversed but there is an edge (y, z) that has not been traversed, proceed along any such edge.

            C. if every edge (y, z) has been traversed, terminate the algorithm.

We will not give a proof here that this algorithm works, but, in fact, it can be shown not only that it works but that it is compatible with our proof of Theorem 2. Other algorithms also exist, including one proposed by Lana Holden early in our study of mathematics and blackwork.

2.3 Eulerian Multigraphs

You may be wondering whether it's necessary for the two threads along each undirected edge to travel in opposite directions, as long as they lie on opposite sides of the fabric. After all, in most circumstances it's impossible to tell by looking which direction a thread was stitched in. Clearly relaxing this restriction doesn't gain us anything, because we can already stitch any pattern. But what if we require that the stitches *always* go in the *same* direction? We could call this "stitching unidirectionally." (We will assume that we are still requiring "reversibility" in the sense that the threads lying on the back of the fabric are in the same places as those in the front.)

In order to keep track of multiple edges going in the same direction, we need something a little different than a graph. A (loop-free) *multigraph* is a set of vertices, V, and a multiset of edges, E, where each edge is an unordered pair of distinct vertices. A pair of vertices has *multiplicity n* if it is connected by exactly n edges. If n is a positive integer, a multigraph has *uniform multiplicity n* (or is *n-uniform*) if every pair of vertices that is connected by an edge is connected by exactly n edges.

Figure 11. A multigraph.

The idea is that a multigraph is just like a graph except that, since a multiset has more than one indistinguishable copy of each element, a pair of vertices in a multigraph may be connected by more than one indistinguishable edge, as in Figure 11. (In this context, our original graphs are sometimes referred to as "simple" graphs.) We could now think of a blackwork pattern as a 2-uniform multigraph, letting us keep track of the number of times each edge has been stitched. (Hopefully, that will come out to be once for the front of the fabric and once for the back!)

A graph may be associated with any multigraph by simply forgetting about any edges connecting each pair of vertices beyond the first. As in the situation with digraphs, a graph may come from any number of different multigraphs. We can uniquely associate a multigraph of uniform multiplicity n to a graph by including in the multigraph n edges connecting each pair of vertices connected by an edge in the original graph, as indicated in Figure 12.

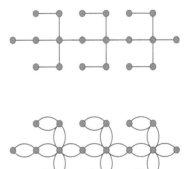

Figure 12. A graph (top) and its associated multigraph of uniform multiplicity 2 (bottom).

You may have noticed that we have not assigned any directions to the edges in our multigraph; it turns out that the proof of our main result on stitching unidirectionally (Theorem 4) is easier that way. This does, however, require a few modifications to some of our earlier definitions. (These are all still quite standard.) A *walk* on a multigraph is a finite alternating sequence of vertices and edges $x_0, e_1, x_1, \ldots, e_n, x_n$, where each $e_i = \{x_{i-1}, x_i\}$. (This is not really different from a walk on a graph.) A multigraph is *connected* if there is a walk between any two vertices.

Observation 4 If a multigraph is connected then the associated graph is connected.

Observation 5 If a graph is connected then the associated n-uniform multigraph is connected.

We still need to make sure that we don't stitch an edge too many times, so we define a *trail* in a multigraph to be a walk where no edge is repeated more times than the multiplicity of that edge. A *circuit* in a multigraph is a trail where $x_0 = x_n$. The *length* of a walk, trail, or circuit is the number of (not necessarily distinct) edges in it. As before, if the length is zero, then the walk, trail, or circuit is called *trivial*. Finally, an *Eulerian circuit* of a multigraph is a circuit that includes all of the edges of the multigraph exactly as many times as the multiplicity of each, and a multigraph is *Eulerian* if it has an Eulerian circuit.

If we can stitch a blackwork pattern unidirectionally (with one thread) then the multigraph must be Eulerian. (Again in this case, being Eulerian implies being connected.) As in the case of directed graphs, it is well known which multigraphs are Eulerian—this actually is the original Seven Bridges of Königsberg Theorem. (Only the "only if" part was rigorously proved by Euler, although he gave a heuristic reason why the "if" part should be true. A valid proof of the "if" part had to wait for Hierholzer over a hundred years later! See [9] for more details and references.) For each vertex v of a multigraph, the *degree* of v is the number of (not necessarily distinct) edges with v as one endpoint.

Theorem 3 (Euler (1736); Hierholzer (1873)). *A multigraph is Eulerian if and only if it is connected and the degree of each vertex is even.*

2.4 Aragonian Graphs

So now we just need to keep track of the direction of the stitching and whether the threads lie on the correct side of the fabric. The side of the fabric is kept track of with parity, as in the Holbeinian case. If $x_0, e_1, x_1, \ldots, e_n, x_n$ is a trail on a multigraph, we say that the *parity* of each edge e_i is the parity of i. (Technically, this definition is an abuse of notation since, for example, we might have $e_1 = e_4 = \{v, w\}$, which in one case has odd parity and in the other case has even parity. In practice, this shouldn't present a problem.)

The next two definitions are again analogous with the Eulerian and Holbeinian case.

Definition 3 An *Aragonian circuit* of a multigraph of uniform multiplicity 2 is an Eulerian circuit where the two edges $\{x, y\}$ are traversed in the order $x, \{x, y\}, y$ and are traversed with opposite parities.

Definition 4 A multigraph of multiplicity 2 is *Aragonian* if it has an Aragonian circuit.

A blackwork pattern can be stitched reversibly with one thread with stitches on the front and back going in the same direction if and only if it is Aragonian. (As before, the requirement of opposite parity corresponds to stitching on both sides of the fabric.) We can characterize Aragonian multigraphs almost as easily as Holbeinian ones. (You might think about how to do this before reading on. The author's first attempt produced conditions that were considerably stronger than necessary!)

Theorem 4 *A multigraph of uniform multiplicity 2 is Aragonian if and only if it is connected, it has a circuit of odd length, and every vertex has degree divisible by 4.*

Corollary 4 *A multigraph of uniform multiplicity 2 is Aragonian if and only if the associated graph is connected, it has a circuit of odd length, and every vertex has even degree.*

Corollary 5 *A multigraph of uniform multiplicity 2 is Aragonian if and only if the associated graph is Eulerian and has a circuit of odd length.*

This time, the proof is virtually the same as that for Eulerian circuits on undirected graphs (again, see [3, Sec. 11.3] or [5, Chap. 7]), with the addition of parity.

Proof of Theorem 4: Let G be the multigraph. If G has an Aragonian circuit, then it must pass through every vertex. Thus for any two vertices, there is a walk from one to the other, so G is connected.

For any vertex v of G other than the starting vertex, each time the circuit comes to v it must depart from v as well. Thus the circuit has traversed two (new) edges with endpoint v. (These edges must be distinct, since if the circuit leaves v by the same edge on which it just arrived then that edge will be traversed in opposite directions.) As the circuit must pass through v twice for each edge, the degree of v must be a multiple of 4. (For the starting vertex, consider the first and last edges of the circuit.)

To show that G has a circuit of odd length, consider the first edge $\{x,y\}$ to appear twice in the Aragonian circuit. Since it must appear with opposite parities, the part of the circuit that includes the first occurrence of $\{x,y\}$ and stops just short of the second occurrence of $\{x,y\}$ will be a circuit of odd length.

Now assume that G is connected, has a circuit of odd length, and every vertex has degree divisible by 4. Let G^s be the graph associated to the multigraph G. Let V be the vertex set of G^s and E the edge set. If E has only three distinct edges $\{x,y\}$, $\{y,z\}$, $\{z,x\}$, then

$$x, \{x,y\}, y, \{y,z\}, z, \{z,x\}, x, \{x,y\}, y, \{y,z\}, z, \{z,x\}, x$$

is an Aragonian circuit of G.

We will proceed by induction, supposing that the result is true if E has fewer than n edges. Consider a G^s where E has n edges. We know that G has a circuit of odd length; call it C. We will first prove that G^s also has a circuit of odd length. Without loss of generality, assume that C has the smallest length of any odd circuit of G. If C does not repeat any edges, then C is also a circuit of G^s so we are done.

Now suppose that C does repeat an edge, so that it is of the form

$$x_0, \{x_0,x_1\}, x_1, \ldots, x_i, \{x_i, x_{i+1}\}, x_{i+1}, \ldots, x_i, \{x_i, x_{i+1}\}, x_{i+1}, \ldots, x_0.$$

Then C can be divided into two smaller circuits,

$$x_i, \{x_i, x_{i+1}\}, x_{i+1}, \ldots, x_i$$

and

$$x_i, \{x_i, x_{i+1}\}, x_{i+1}, \ldots, x_0, \{x_0, x_1\}, x_1, \ldots, x_i.$$

But since C has odd length, one of these smaller circuits must also have odd length. This contradicts the fact that C has the smallest length of any odd circuit of G.

Therefore, we can assume that C is a circuit of G^s. If C contains a copy of every edge in E, then we are done; simply follow the circuit C twice. If not, let K be the sub-multigraph of G obtained by removing all copies of the edges in C with associated graph K^s. Note that the degree of a vertex in G^s is half the degree of the same vertex in G. Therefore, because we remove edges two copies at a time, each vertex of K has degree divisible by 4 and each vertex of K^s has even degree. Now K^s may not still be connected. However, each component of K^s is connected (by definition) and has an Eulerian circuit, since each vertex in the associated graph has even degree. (This circuit may be trivial if the component only has one vertex.)

We construct an Aragonian circuit on G by travelling along the circuit C with detours as follows: Each vertex w of C is part of some component L^s of K^s. Each vertex of L^s has even degree, so it has an Euler circuit (possibly a trivial one). If L^s has an odd number of edges, we traverse the Euler circuit of L^s twice. Because traversing L^s once switches the parity, this produces an Aragonian circuit of L. If L^s has an even number of edges, we traverse the Euler circuit of L^s once. In either case, the detour is of even length, so it doesn't affect the parity of the edges of C.

When we arrive back at the first edge of C, we have switched parities, since C has an odd number of edges. We follow C again, this time only detouring onto L^s if it has an even number of edges and traversing its Euler circuit once. Note that the edges of L^s have opposite parities from the first traversal of L^s. Because we have traversed each edge of G^s twice with opposing parities, we have completed an Aragonian circuit of G.

3   Teaching Ideas

Although students typically don't encounter formal graph theory until a college course such as "Finite Mathematics" or "Discrete Mathematics," there is really nothing in the basic ideas that is beyond an elementary school student. Dots, lines, and arrows are the basic objects, and the ideas of "paths" and "walks" using these objects are extremely natural. (In fact, there is a type of graph known as a *dessin d'enfant*, or "child's drawing," with deep connections to important areas of mathematics!) Likewise, kits introducing children to embroidery techniques are aimed at children as young as five or six years old. These kits usually have designs stamped onto plastic canvas, which the child stitches using three- or four-ply yarn and large blunt plastic needles according to the stamped design. (The sort of pattern I've been describing, with written directions and figures, is usually called a "counted" design.) Preparing a stamped blackwork project would be more work for a classroom teacher than a counted one but would still include much of the mathematical value. Blank pieces of plastic canvas can be purchased in a variety of sizes (from 4″ by 4″ to 8.5″ by 11″ and beyond) and marked with a permanent marker. Figure 5 would be an excellent place to start with younger children. A second project might be for students to design their own patterns, mark them on plastic canvas, and try to figure out how to stitch them with one continuous strand.

Counted blackwork patterns would probably be more suitable for slightly older children. They could also start with Figure 5 and proceed to designing their own patterns. The "detail graphs" from Section 4, Figures 16 and 17, could also be done as stand-alone patterns. High-school or college students could work their way up to designing their own patterns and determining the orders in which they should be stitched using the method from Section 1.3.

A more exploratory activity could proceed as follows: Describe to the class the rules for doing blackwork. Then give the class a blackwork pattern and suggest that the students stitch it with the challenge that they use as few pieces of thread as possible and that they keep track of how they completed the stitching sequence. Have the students work in pairs, so that one student is doing the stitching while the other is keeping track of the stitching sequence. After a sufficient period of time, have a discussion about the students' results, strategies, conjectures, and so forth. This activity should prompt students to conjecture Theorem 4 and can be modified to deal with unidirectional stitching as well.

So what is the mathematical value? First, the simple idea that graph theory is part of mathematics is useful to counteract the idea that math is only about numbers. (Despite much progress in introducing other ideas, such as symmetry, into the early curriculum, this idea is still very common!) A first step for young children might be to introduce the ideas of graph theory informally and then ask them to "translate" the question of whether a design can be stitched into graph-theoretic language. Perhaps some speculation on the answer would be in order!

For slightly older children, an informal version of the proof could be presented, especially after the students have some experience making and stitching their own patterns. (This might be a suitable activity for a math club or other extracurricular math activity, rather than trying to fit it into a standard high-school curriculum.)

Alternatively, students might be presented with only the Eulerian version of the proof and asked how to modify it for the Holbeinian version.

In addition to the ideas of graphs and proofs, two other very important ideas are contained in this chapter, namely the idea of an algorithm for doing something and the idea of recursion. A teacher or club leader might describe the idea of using an algorithm before presenting the one given here. Then the participants could try to come up with their own.

For students in an upper-level high-school or college mathematics course, the full mathematical definitions could be presented, of course, using the stitching as a motivating factor. Students who are squeamish about mathematics, and even some who aren't, might find having a practical hands-on application of graph theory to be very refreshing if they have not seen one before. Students should be encouraged to think about alternate ways to prove the theorems given. (There are many known!) Also, there are many possible extensions of the results, such as requiring each edge to be

traversed more than twice, or requiring some edges to be stitched in a "Holbeinian" and others in an "Aragonian" fashion. Finally, students with a strong interest in computer programming should be encouraged to think about how to program a computer to stitch a pattern reversibly, using the iterative algorithm from Section 2, a recursive algorithm based on the proof given, or some other algorithm. (There are many choices!)

4 How to Embroider a Holbeinian Graph

Materials

The following pattern (shown front and back in Figures 13 and 14) may be stitched on any evenweave fabric, such as linen or Aida cloth. I like 32 count linen, but stitchers who are less nearsighted tell me they prefer a looser weave. If you use Aida cloth, note that because of the construction you will be stitching in every hole rather than every other hole. (Figure 2 and the

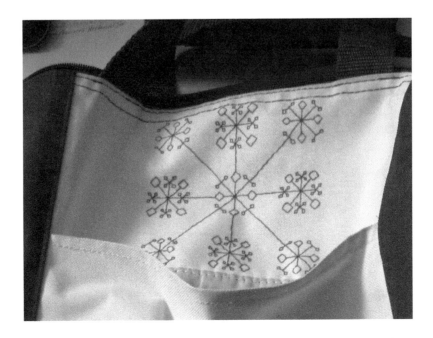

Figure 13. The Embroidered Holbeinian Graph can be stitched onto a tote bag.

Figure 14. And what is the preferred bag? One from the Joint Mathematics Meetings, of course!

subsequent figures show every other hole.) The pattern is 64 stitches by 64 stitches, which will be 4 ″ by 4″ on 32 count linen. Any contrasting color of embroidery thread may be chosen, and there are lots of easily available choices. I would use silk or one strand of stranded cotton, but if you prefer rayon or metallic (e.g., "goldwork"), by all means try it. At the 4″ by 4″ size you will only need a yard to a yard and a half of thread, depending on how efficient you are with your thread.

How to Read the Diagrams

Note that in the spirit of recursion (as described in Sections 2 and 3), we have provided a schematic graph (Figure 15) and two detailed graphs (Figures 16 and 17), in addition to the complete graph (Figure 18).

One way of stitching the pattern in accordance with the method we have given would be to start at the center of Figure 15 and go up the northeast spoke toward section A. After going out and back on the northeast spoke, go on to the east spoke and so on clockwise around the graph.

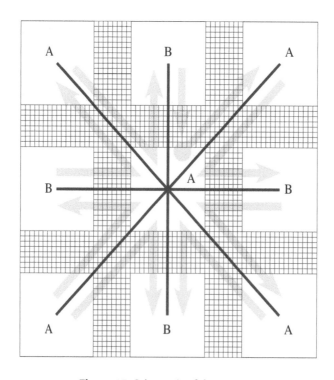

Figure 15. Schematic of the pattern.

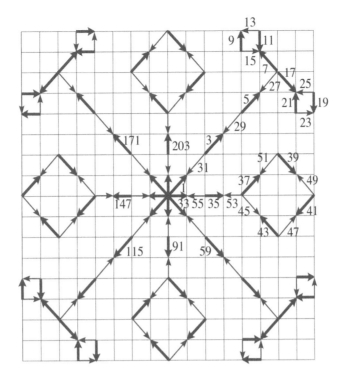

Figure 16. Detail section A of the pattern.

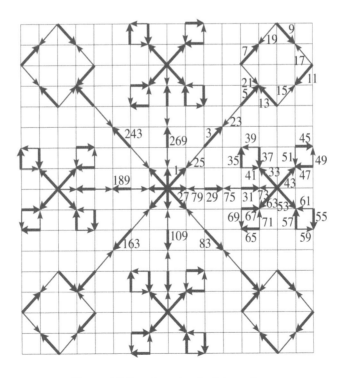

Figure 17. Detail section B of the pattern.

Figure 18. The complete pattern.

In more detail, the stitching would follow this procedure:

* Start at the center of the pattern.

* Do stitches 1–26 of section A.

* Keep going northeast to the center of the northeast section A.

* Stitch section A.

* Return to stitch 27 of the center section A.

* Do stitches 27–48 of section A.

* Keep going east to the center of the east section B.

* Stitch section B.

* Return to stitch 49 of the center section A.

* …and so on.

Numberings have only been given once for each detailed section of the pattern, and complete numberings have been given only once for each subsection of the detailed section. (Even those aren't quite complete, since only the odd numbers are given, but the arrows should make the direction of stitching clear. Heavy arrows indicate the "outward" or "running" path and lighter arrows are the "return" or "double running" path that finishes the stitch.) A stitcher should use this information in each different place where the subsection appears.

Happy stitching!

Bibliography

[1] Barnett, Lesley. *Blackwork*. Search Press, Tunbridge Wells, UK, 1996.

[2] Drysdale, Rosemary. *The Art of Blackwork Embroidery*. Charles Scribner's Sons, New York, 1975.

[3] Grimaldi, Ralph P. *Discrete and Combinatorial Mathematics*, Fifth Edition. Addison Wesley, Reading, MA, 2003.

[4] Gross, Jonathan L., and Yellen, Jay. *Handbook of Graph Theory*. Discrete Mathematics and its Applications. CRC Press, Boca Raton, 2004.

[5] Harary, Frank. *Graph Theory*. Addison Wesley, Reading, MA, 1969.

[6] Holbein, Hans, the Younger. *Jane Seymour, Queen of England*. Oil on wood, 65.5 x 40.5 cm, 1536, held by Kunsthistorisches Museum, Vienna. http://www.wga.hu/frames-e.html?/html/h/holbein/hans_y/1535h/02seymou.html

[7] König, Dénes. *Theory of Finite and Infinite Graphs* (translated from the German by Richard McCoart; with a commentary by W. T. Tutte and a biographical sketch by T. Gallai). Birkhäuser, Boston, 1990.

[8] Scoular, Marion E. *Why Call It Blackwork?* Marion E. Scoular Sherwood Studio, Duluth, GA, 1993.

[9] Wilson, Robin J. *History of Graph Theory*. In [4], Section 1.3, pp. 29–49.

CHAPTER 10

stop those pants!

SARAH-MARIE BELCASTRO
CAROLYN YACKEL

1   Overview

What do a stop sign and a pair of pants have in common? "Pardon me," you might say, "but my pants are neither metallic nor red." Over the vehement protests of those of who sew pants, the answer is they can both be made from octagons. This strange observation raises two important questions. If pants can be made from octagons, then why don't tailors make pants from octagonal patterns? And, if tailors don't make pants from octagonal patterns, then why did anyone ever think that pants could be made from octagons?

Let's consider the second question first. What would make anyone think to try to make a pair of pants out of an octagon? The key to the answer lies in Chapters 1, 4, and 7. In each of these chapters, the author explains a topological surface. She represents the surface by drawing a rectangle and discussing how to glue together the sides. However, not every topological surface can be represented by a rectangle. For example, a two-holed torus can be represented by gluing together

corresponding sides of an octagon, as shown in Figure 1, and cannot be represented by a polygon with fewer than eight sides.

Now, it is easy to see how a creative mind might come up with other ways to glue an octagon into a surface. Even if one sticks with gluing together all pairs of sides, there are a number of possible outcomes (which the reader is encouraged to explore on her own). The gluing that produces a pair of pants neglects glue on four of the sides, which allows for the waist and leg openings as in Figure 2.

Experiment 1 Duplicate Figure 2 and cut out the octagon. Tape the identified sides together (the sides that are supposed to be sewn), matching direction arrows as shown. First attach the sides labeled a to each other, then attach the b sides to each other. Notice that the result has three holes. One of the holes is formed by two of the original sides of the octagon coming together for the waist, and the other two holes are each formed by one side curling around to make a leg hole. With imagination, these are pants.

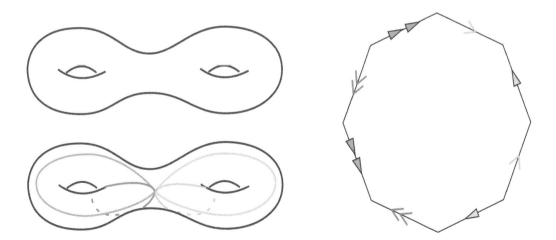

Figure 1. The two-holed torus (top left) together with its octagonal representation (right) and the corresponding fundamental cycles marked (bottom left).

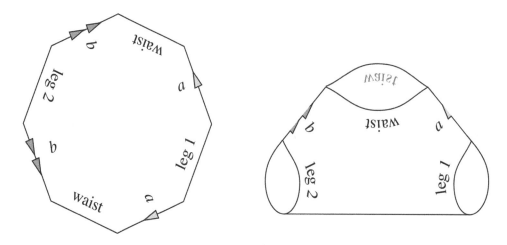

Figure 2. An octagon that folds up into a pair of pants.

A little paper folding helps one to visualize this octagon folded in half as pants. Fold the center point of the octagon a bit more than half way up to the waist as shown in Figure 3(a), and crease the middle third of this fold horizontally. (Each lettered arrow in the figure indicates an action transforming one picture to the next.) Then fold the legs down from the center of the octagon, as in Figure 3(b). This will cause the sides of the paper to bend up—don't worry—allow this to happen. Let the points of the waist edges meet in the center, and then crease what would be the upper thighs if these really were pants, as Figure 3(c) shows. Now look at the shape of the pants (the back is shown with Figure 3(d)).

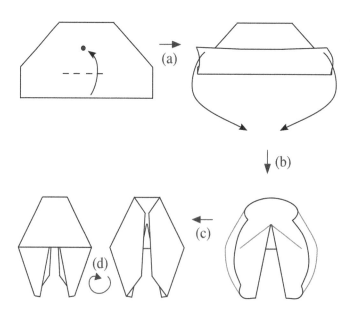

Figure 3. Making the octagon look more like a pair of pants.

Those of you who have been tenaciously shopping at world clothing stores located in college towns will recognize this style, sold as Thai fisherman pants. However, even those pants do not have *such* a gathering of material at the crotch (where the inside seams of regular pants meet). We would rather have a result more like Thai wrap pants, which are loose but have no excess crotch fabric.

Topologically (that is, in terms of the number of holes), we have successfully created pants. However, geometrically (that is, in terms of proportions, curviness, and distances), we have more work to do. Furthermore, our model shows that the problem is not the sizing but the flatness of the original octagon.

1.1 The Curvature of Wearable Pants

In order to avoid flatness, we will work with curved surfaces. Positive constant curvature yields a sphere, and negative constant curvature yields a hyperbolic plane or pseudosphere; it is not easy to make any of these surfaces accurately from fiber. We can approximate them well, which we will do with knitting in Section 4. For now, we will approximate them poorly with paper— even a poor approximation, however, will serve to illustrate the true nature of pants.

Our paper approximation will consist of a set of triangles glued together. While each piece is flat, the object made from them will not be. Mathematicians say that such a surface is locally flat, but globally curved. The global curvature will arise from a change in the angles around certain points in the surface; ordinarily, adding the angles around a point in a piece of paper gives 360°, but in the next two experiments this will not be the case.

Experiment 2 (a) Figure 4 shows eight triangles that can be made into an octagon with positive average curvature comparable to that of a unit sphere. First, duplicate Figure 4 and cut out the shape. Crease along the edges between the triangles, then tape the arrowed

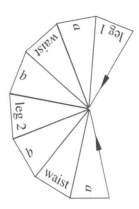

Figure 4. Triangles that form a spherical octagon that folds into a pair of pants.

sides together. Now, just as in Experiment 1, attach the *a* sides together and attach the *b* sides together. As before, the result has three holes, corresponding to the waist and ankles of a pair of pants. You can make the resulting shape look more pants-like by following the same folding sequence as given in Figure 3. At this point, you'll notice that material has been added in exactly the wrong region of the pants—there is even more material at the crotch! (To understand why this happens, read Section 2.)

Experiment 2 (b) Obtain a close-fitting knitted hat (or make one, using the instructions in Chapter 2). It has curvature similar to that of a sphere. Mark eight points along the brim with thread as follows: Choose any point on the brim as the first mark. Fold the hat in half at this mark, and mark the point opposite (at the other end of the fold-line). Bring the two marked points to each other; this makes two more folds that indicate the next two points to be marked. Similarly, mark the points halfway between each of the four adjacent pairs of points. Label the resulting segments in order: waist, *a*, leg 1, *a*, waist, *b*, leg 2, *b*. Baste together the *a* segments and also the *b* segments. Notice that these are really terrible pants, with much too much fabric in the crotch area.

If we are going to manage to construct reasonable pants, they will have to come from a surface with negative curvature.

Experiment 3 Figure 5 shows four pairs of triangles that can be made into an octagon with negative average curvature. First, duplicate the figure and cut out the four shapes. Tape the arrowed sides together. Now, just as in Experiment 1, attach the *a* sides together and attach the *b* sides together. As before, the result has three holes, corresponding to the waist and ankles of a pair of pants. But, unlike the previous two experiments, this actually looks like a pair of pants!

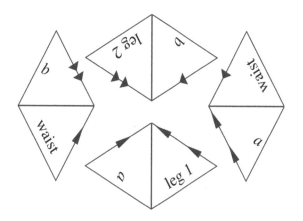

Figure 5. Triangles that form a hyperbolic octagon that folds up into a pair of pants.

In reality, pants designers do not use uniform curvature when making patterns. (This is why commercially available patterns do not use octagons.) They want zero curvature along the legs of the pants and zero curvature near the waist—but negative curvature near the crotch. Such a pattern for Thai wrap pants is shown in Figure 6, where you may note that only the crotch is curved.

In the remainder of this chapter, we will explain the mathematics of curved surfaces in detail, give ideas for how this mathematics can be used in the classroom, and end with a pattern for a uniformly curved pair of pants. These pants are curved all over, not just in the crotch,

and are sized for classroom demonstration (or for a human baby).

Figure 6. A pattern for Thai wrap pants. Duplicate twice and attach the copies along the curved edge to make actual pants.

2 Mathematics

Paradoxically, all curves are flat. That is because any curve in space ($\mathbb{R}^n$) is parametrized by a single variable and as such is the image of $\mathbb{R}^1$ under an embedding map. While every curve is a line embedded into space, surfaces have intrinsic curvature, that is, curvature that is measurable without any knowledge of an embedding into space. This is known as Gaussian curvature because, well, it was Gauss' idea back in the 1800s.

2.1 Intrinsic Curvature vs. Extrinsic Curvature

We define intrinsic curvature locally (though not as locally as in the definition of a manifold!) by examining a circle drawn on the surface around a point p. If the circle has circumference equal to $2\pi r$, then the Gaussian curvature is zero and the surface is locally flat. If it has circumference greater than $2\pi r$, the curvature is negative; if the circumference is less than $2\pi r$, the curvature is positive. One way to think of this is to remove the circle (and the interior) from the surface where it lives, snip the circle to its center, and lay it down on a plane [3]. If the circle is still closed and doesn't overlap itself, then we have flatness. If the circle is open, the positive curvature may be measured by the angle missing; if the circle overlaps itself, the negative curvature may be measured by the overlap angle.

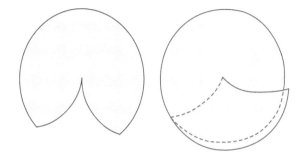

Figure 7. Discs from positively (left) and negatively (right) curved surfaces placed on a plane.

To be more precise, denote the angle defect (or excess) of such a circle by θ_r, as the angle is dependent on the radius of the circle (see Figure 7). We define the *intrinsic curvature at a point* of a surface to be

$$K_p = \lim_{r \to 0} \frac{\theta_r}{A_r},$$

where A_r is the area of the circle. Note that if our surface is not flat, then $A_r \neq \pi r^2$. In practice one might wish to measure θ using *parallel transport*—that is, move a vector around a square with a corner at our point as in [6], because this allows one to approximate the area easily. We can even generalize the definition of K_p to any closed curve around a point p on our surface S. Consider a disk D with radius r and boundary ∂D in $\mathbb{R}^2$, and consider the image of this disk under a smooth map $\gamma : \mathbb{R}^2 \to S$. The angle defect (excess) from parallel transport of a vector around ∂D is given by $\int_D \gamma^*(K \, dvol)$, where $dvol$ is a volume element in S and γ^* is the pullback of γ (see [6, p. 171]). One can then compute K_p from the $\lim_{r \to 0}(\int_D \gamma^*(K \, dvol))/(\int_D \gamma^*(dvol))$.

In the world of fiber arts, it would be rare to encounter a surface not embedded in $\mathbb{R}^3$. Thus it is useful to consider extrinsic measures of curvature as well.

One extrinsic measure of curvature at a point p is $K_p = 1/(r_1 r_2)$, where the r_i are the radii of curvature of curves formed by extending local coordinate vectors to curves on the surface [5]. This is equivalent to another extrinsic measure, where we define K_p to be the determinant of the differential of the Gauss map [2]. Because any differential is a linear operator, it makes sense to express it as a matrix; when we choose suitable (orthonormal) coordinates, we obtain

$$K_p = \det \begin{pmatrix} 1/r_1 & 0 \\ 0 & 1/r_2 \end{pmatrix},$$

which again produces $K_p = 1/(r_1 r_2)$.

Consistency requires that we reconcile the intrinsic and extrinsic measures of curvature, and there are several ways to do this. One is Gauss' Theorem Egregium, which states that intrinsic and extrinsic curvature are equivalent, and another is Gauss' Formula [2], which reduces an extrinsic expression for K to an intrinsic expression for K. We will find the Gauss-Bonnet Theorem more useful.

The usual formulation of the Gauss-Bonnet Theorem is for compact orientable surfaces S and states that

$$\frac{1}{2\pi} \int_S K \, dvol = \chi(S).$$

Here $\chi(S)$ is the Euler characteristic as defined in Chapter 1. This means that a change in the embedding of a surface that changes the curvature in one spot must also change the curvature elsewhere, to "cancel out" the original curvature change. The theorem can be proven starting with a definition of K as extrinsic curvature, or starting with a definition of K as intrinsic curvature, as in [6, Section 7.3]. Thus the two notions of curvature are compatible, so a global measure of K taken extrinsically may be interpreted intrinsically and vice versa.

In practice, this means that Gaussian curvature is invariant under isometric mappings. In other words, throwing a knitted object across the room (thus changing the embedding of the surface) is a local isometry because nothing stretches and the overall curvature does not change.

For the purposes of this chapter, we will only consider surfaces of uniform constant curvature, so that K is independent of p. Using the extrinsic definitions, we will have $r_1 = \pm r_2$ so that we may speak of $K = 1/r^2$ in the positive curvature case and $K = -1/r^2$ in the negative curvature case. The astute reader will now object that Gauss-Bonnet is irrelevant for surfaces of uniform negative curvature as they are not compact, to which we respond that a version of Gauss-Bonnet for simplices will come to our aid in Section 2.2.

2.2   Discrete Measures of Curvature

Our goal in this chapter is to show how to create uniform constant negative curvature by knitting. Knit stitches are distinct and thus not continuous, so we will need to measure curvature discretely. Furthermore, the definitions of curvature we gave earlier do not make sense in discrete situations. An attempt to measure K intrinsically near a polyhedral vertex fails because the angle excess is constant while the local area goes to 0; a similar extrinsic attempt fails because one cannot orient local coordinate vectors at a singular point.

More to our purposes, there is a local formulation of Gauss-Bonnet, so we are not restricted to compact surfaces. Consider a 2-simplex Δ with boundary $\partial\Delta$ in $\mathbb{R}^2$ and the image of this 2-simplex under a smooth map $\gamma : \mathbb{R}^2 \to S$. The angle defect (excess) from parallel transport of a vector around $\partial\Delta$ is again given by $\int_\Delta \gamma^*(K \, d\mathrm{vol})$, and Gauss-Bonnet for simplices says that $\int_\Delta \gamma^*(K \, d\mathrm{vol})$ = the integral of geodesic curvature over $\partial\Delta$ with respect to arc length $+ \, 2\pi - \sum_i \alpha_i$, where the α_i are exterior angles of $\gamma\Delta$. (Here an exterior angle is the supplement of an interior angle.) In fact, we need not be restricted to simplices; we may think of any polygon as composed of simplices, and in computing local

Gauss-Bonnet the integrals of the interior segments cancel. We then only need to deal with the perimeter of the polygon.

Of course, we do not really want to calculate the integral of geodesic curvature with respect to arc length or even think about what that means. Luckily, we will not have to. We will restrict ourselves to polyhedral approximations to curved surfaces, so that the surfaces we consider will be piecewise flat. Therefore, the geodesic curvature of any polyhedral boundary with respect to arc length will be zero, and we may ignore this term in the Gauss-Bonnet formula.

To see how this works, let's do a sample calculation based on the actual pants of Figure 6. We want to calculate the curvature in one of the "corners" of the crotch, and we will do so by approximating the surface polyhedrally as shown in Figure 8.

The neighborhood N of the vertex that hosts the crotch curvature has six $45°$-$45°$-$90°$ triangles, and thus each interior angle is $90°$ (π radians). This means that each exterior angle is also $90°$ (the surrounding fabric is flat), so that

$$\int_N \gamma^*(K \, d\mathrm{vol}) = 2\pi - \sum_i \alpha_i = -\pi.$$

Because K is constant on our vertex and γ is piecewise the identity on N,

$$\int_N \gamma^*(K \, d\mathrm{vol}) = K \int_N \gamma^*(d\mathrm{vol}) = K \int_N d\mathrm{vol} = K \, \mathrm{Area}(N).$$

Notice that this is consistent with our earlier definition of curvature at a point:

$$K = \lim_{r \to 0} \frac{\int_N \gamma^*(K \, d\mathrm{vol})}{\int_N \gamma^*(d\mathrm{vol})}$$

$$= \lim_{r \to 0} \frac{K \, \mathrm{Area}(N)}{\mathrm{Area}(N)} = K.$$

We may scale our neighborhood so that $\mathrm{Area}(N) = 1$ and $K = -\pi$.

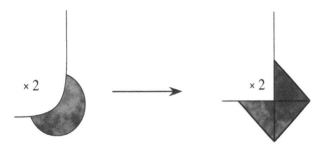

Figure 8. A local patch of the Thai wrap pants crotch (left) and its polyhedral approximation (right).

To complete our analysis, we must consider curvature calculations for pieces of a surface that have more than one vertex. For example, consider a polyhedral surface comprised of equilateral triangles, where some vertices are of order six and some are of order five (as in, for example, a geodesic dome). This will have positive curvature. A surface comprised of squares, with some vertices of order four and some of order five, will have negative curvature. Because we are considering only uniformly curved surfaces, we know that any surface we consider can be partitioned into regions of equal area, such that

* each region contains the same number of internal vertices,

* the degree sequence of the internal vertices is the same for each region, and

* every vertex on the boundary of each region is flat, i.e., is surrounded by angle 2π.

(The polygons comprising the surface may need to be subdivided in order to achieve the last point.) Then, K for the surface will be the same as K for each region.

2.3 Uniform Curvature in Knitting

To compute discrete curvature for a knitted fabric, we will use a polyhedral model. Knit stitches, ignoring increases or decreases, form a rectangular grid (with one stitch sitting inside each rectangle; see Figure 9). Furthermore, an individual knit stitch is flat—it has no Gaussian curvature, geodesic curvature, or any other kind of curvature. Thus for a single stitch there is no contribution from the integral of the geodesic curvature, as the integral of 0 with respect to arc length is still 0.

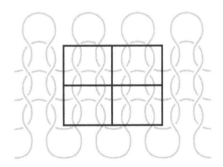

Figure 9. A section of knitting (left) and the same section with a rectangular grid over it (right).

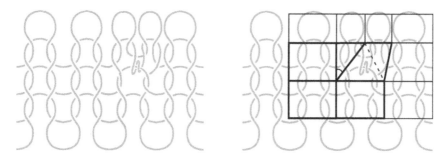

Figure 10. An increase stitch (left) shown with the resulting change in the polygonal mesh (right).

And the sum of the exterior angles of a stitch is 2π, which yields an angle defect of 0, and thus $K = 0$.

If we look at a piece of stockinette (plain knitted fabric), it likewise has no curvature, as each vertex of the rectangular grid has no angle defect and each surrounding polygon is flat. We will see that curvature of knitted fabric arises from alterations of this rectangular grid.

Garment shaping in knitting is produced using *increases* and *decreases*. True to their names, they are types of stitches that respectively increase and decrease the number of stitches in a row. Figures 10 and 11 show yarn-flow schematics of increase and decrease stitches. Ordinarily we have four stitches surrounding one vertex of our rectangular grid. In the case of an increase, we create two stitches from one, or add a stitch between two other stitches, so that there are five stitches around one vertex. Each stitch is effectively a rectangle in the underlying polygonal mesh, and making an increase can be interpreted in terms of that mesh by having two rectangles grow out of the top of one rectangle. The lower edges of the two new rectangles are identified (i.e., have been glued together), but the upper edges are separate; thus, the rectangles overlap, and the result is an extra triangle added to the mesh originating at the five-stitch vertex. See Figure 10 for a depiction of the mesh atop an increased stitch. In reality, every stitch would be the same size as every other stitch, though the lower halves of the two stitches in the increase stitch would be compressed.

Interestingly enough, making a decrease has the same effect on the mesh. Intuitively, a decrease stitch looks like an increase stitch if the fabric is held upside down. More technically, Figure 11 shows two mesh rectangles with separate lower edges and upper edges

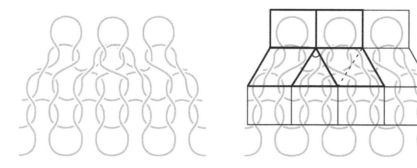

Figure 11. A decrease stitch (left) shown with the resulting change in the polygonal mesh (right).

identified—this is literally the way a decrease stitch is constructed, as seen in the left-hand portion of the diagram.

Of course, these diagrams are not drawn to scale, so the apparent angles are not reflective of actual knitted fabric. We will give the technical mathematical details of the increase and decrease constructions in a future publication.

Now, let us revisit the constraints we listed at the end of Section 2.2 in the context of knitting a surface of uniform negative curvature. We want to design a pattern that will allow us to subdivide the surface into regions of equal area and curvature such that every vertex on the boundary of each region contributes zero curvature. The simplest way to accomplish this is to include exactly one decrease in every sequence of n stitches for some fixed $n \geq 2$. (For any knitters reading this far, this means *K($n - 2$) K2tog*.) This is the inverse construction to that given in [4]. A diagram of part of a surface subdivision is shown in Figure 12, where each region is composed of n blue boxes, one of which contains an excess-curvature vertex associated with a decrease stitch.

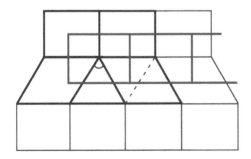

Figure 12. Subdividing the polyhedral model of a uniformly negatively curved knitted surface.

This scheme clearly partitions the surface into regions of equal area and curvature. It remains to be seen that every boundary vertex contributes no curvature; this is only in question for the two lower vertices adjacent to the excess-curvature vertex. We may subdivide the extra triangle into lower and upper sections, so that the lower section is a rectangle and the upper section is a triangle with right angles. (Examination of knitted fabric shows that this is a reasonable and accurate subdivision.) The discerning reader will notice that this not only assures that the two questionable vertices induce no curvature, but also regularizes the two vertices directly below—these are contained in another region, and to keep the curvature uniform we want only decrease-stitch vertices to contribute curvature.

It turns out that the angle excess at a negative-curvature vertex is approximately $30°$ (determined from examining actual knitted work). We may think of each stitch as having area 1 (that is, one stitch unit). The corresponding area added at a negative-curvature vertex is thus approximately $1/(2\sqrt{3})$ stitch units. This means that each region has angle excess of $\pi/6$ with area $n+1/(2\sqrt{3})$ stitch units, so using local Gauss-Bonnet we have

$$K_{\text{region}} = \frac{-\pi}{6n + \sqrt{3}}.$$

From this we can calculate the extrinsic measure of the radius of curvature

$$r = \sqrt{\frac{6n + \sqrt{3}}{\pi}},$$

as for uniformly negatively curved surfaces $K = -1/r^2$. We have now obtained a formula for the radius of curvature in terms of the frequency of decrease stitches.

As a practical note, this construction produces two particularly simple surfaces of constant negative curvature. One is a segment of the hyperbolic plane, produced by casting on an unreasonably large number of stitches, using flat knitting (not in the round), and following the regular decreasing procedure until only $n - 1$ stitches are left, then casting off. (This is essentially the same as Daina Taimina's crochet construction [8], but in reverse; she uses increases where we use decreases.) The other is a pseudosphere, made by again casting on

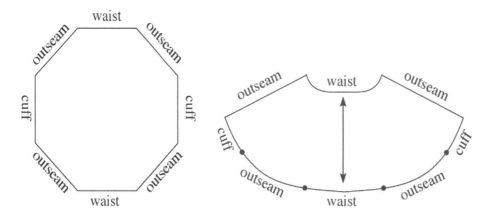

Figure 13. Pants identifications for a flat octagon (left) and a hyperbolic plane section (right).

2.4 Knitting Pattern Construction for Hyperbolic Pants

As discussed earlier, a uniformly curved pair of pants can be made from an octagon. Figure 13 (left) shows the usual identifications for pants, made on a flat octagon. To make pants with uniform negative curvature, we must use a patch from a hyperbolic plane. It may seem at first more natural to use a pseudosphere, but that leaves a hole for a tail or other central appendage; this has limited practical use in our society. A hyperbolic plane patch with pants identifications is shown in Figure 13 (right).

There are some mathematical and knitting-design aspects to the diagram that are worth noting. First, as drawn here our hyperbolic plane patch is not a true octagon because five of the sides are not hyperbolic geodesics. Using this model is somewhat akin in Euclidean space to using an octagon where three of the sides are straight and five are slightly convex. In practice, the difference between these octagon boundary curves and true geodesics is small because pants made

to fit humans have relatively low curvature. (Note that the less curved the space, the smaller the error between a geodesic and a nearby curve.) Second, the double-headed arrow marks the waist-to-waist measurement, which is the same length as the outseam. The inseam lies along a geodesic that runs through the center of the waist-to-waist to the center of a cuff. Thus, hyperbolic pants are determined by the waist, outseam, cuff, and waist-to-waist measurements rather than by the standard waist, inseam, and cuff measurements of sewn pants. Third, the two straight outseams and the waist-to-waist are measured in rows, whereas the other outseams, the cuffs, and the waist are measured in stitches.

Seamstresses are now very worried: most humans have much longer outseams than waist-to-waist measurements. For example, a Misses size 10 has a waist-to-waist of about 23″ and an outseam of about 42″. However, young children have waist-to-waist measurements that are comparable to their outseams. This allows us to use a simple design for baby-sized pants. Adult pant design is more complex, and our patterns will appear in [1].

2.5 Hyperbolic Pant Design for Babies

The first step in any knitted pattern design is to produce a gauge swatch. This gives the number of stitches per

Figure 14. The desired pants.

inch and number of rows per inch for a particular yarn knitted using a particular needle size. We need only the waist-to-waist measurement converted to rows and to stitches, the cuff measurement converted to stitches, and the waist measurement converted to stitches to determine the remainder of the pattern for pants as shown in Figure 14. Let w be the waist measurement in stitches, c the cuff measurement in stitches, r the waist-to-waist measurement in rows, and s the waist-to-waist measurement in stitches. We would like to produce a hyperbolic plane patch that has $w/2 + 2s + 2c$ stitches cast on and $w/2$ stitches bound off. In other words, there are $w/2 + 2s + 2c$ stitches in the first row and $w/2$ in the rth row. We will need to determine the pattern of decreasing, that is, determine $(n - 2)$ in the pattern *K$(n - 2)$, K2tog*.

The number of stitches in the second row of the pattern is

$$\left\lceil (w/2 + 2s + 2c)\left(\frac{n-1}{n}\right)\right\rceil.$$

Assuming that we begin each row with K$(n - 2)$ (or P$(n - 2)$ as appropriate), the number of stitches in the third row of the pattern is

$$\left\lceil \left\lceil (w/2 + 2s + 2c)\left(\frac{n-1}{n}\right)\right\rceil\left(\frac{n-1}{n}\right)\right\rceil,$$

and, if we define

$$f(x) = \left\lceil x \cdot \left(\frac{n-1}{n}\right)\right\rceil,$$

then the number of stitches in the ith row is f applied $(i-1)$ times to $x = w/2 + 2s + 2c$. Fortunately, *Mathematica* has a command Nest that iterates functions, and thus we can define

```
stitchesleft[x_]:= Ceiling[x*((n-1)/(n))]
```

and we know that

$$\text{Nest}[\text{stitchesleft}, w/2 + 2s + 2c, r] = w/2.$$

We can use this equation along with trial-and-error to find n, but it is desirable to have a close starting value for n. We begin by making a continuous approximation. Using the equation $y = y_0 e^{kt}$ for the number of stitches as a function of the row index, we can solve for the approximate stitch ratio $e^k \approx n - 1/(n)$. Renumbering the initial row as zero instead of one, we have two points $(0, w/2 + 2s + 2c)$ and $(r - 1, w/2)$ that (after a couple of lines of algebra) determine

$$e^k = \left(\frac{w/2}{w/2 + 2s + 2c}\right)^{1/(r-1)}.$$

This produces a serviceable pattern: long-tail cast on $w/2 + 2s + 2c$, and alternate row-patterns of *P($n - 2$), P2tog* and *K($n - 2$), K2tog* for r rows; bind off. Joining along the outseams is aided by the placement of split stitch markers at stitches c, $c+s$, $c+s+w/2$, and $c+2s+w/2$ in the cast-on row. However, the decreases will line up near the edges of the knitted fabric, creating lines of locally higher curvature. This effect can be mitigated by changing the pattern to use a cycle of four rows instead of two, as *P($n - 2$), P2tog*, *K($n - 2$), K2tog*, P($n - j$), *P2tog, P($n - 2$)*, and K($n - j$), *K2tog, K($n - 2$)*, where j is a value small compared to n but at least an inch in stitch units.

3 Teaching Ideas

While there are many possible activities that relate to this chapter, in this section we focus on those specifically related to curvature and to pattern creation. Some of the questions and activities are appropriate for elementary students, whereas some require second-year calculus. Be careful to keep the level in mind when using them in a classroom.

3.1 Paper and Tape

This trio of activities explores the differences among the three types of curvature.

⋆ Make the flat version of the octagonal pants outlined in Experiment 1, using Figure 2. Pay specific attention to the "gluing," which will be done with tape. Be sure you understand where the waist of these pants is, and where the legs come out. What's wrong with the pants? Why would they be uncomfortable to wear (other than that they are made of paper)?

⋆ Make the spherical version of the octagonal pants, outlined in Experiment 2(a), using Figure 4. Why are they called spherical? What makes them octagonal? Where's the waist? Where do the legs come out? Are these better or worse than the flat version? Why? Why would they be uncomfortable to wear (other than that they are made of paper)? Or would they fit appropriately?

⋆ Make the hyperbolic version of the octagonal pants, outlined in Experiment 3, using Figure 5. Why are they called hyperbolic? What makes them octagonal? Where's the waist? Where do the legs come out? Are these better or worse than the flat version? Why? Or would they fit appropriately?

3.2 Plastic and Fiber

These activities are particularly useful for advanced secondary and post-secondary geometry classes.

Modeling Constant Positive Curvature. While Section 3.1 used discrete models of curvature, here we make octagonal pants out of a surface that has constant positive curvature. The following reprises Experiment 2(b).

If you have a Lenart sphere available to you, then use it. Begin with a transparency cap on the top of the Lenart sphere and remove it when it is time to glue edges. Otherwise, you may use any flexible half-sphere or almost-half-sphere that you can find. Next, you need to construct or mark off an octagon on the half-sphere, where the eight sides have equal length. Remember that edges on a sphere are segments of great circles (circumferences of the whole sphere), just like the pictures of flight paths shown in airline magazines. Therefore, one easy way to make a regular octagon is to divide the edge of the half-sphere, which is itself a circumference, into eight equal segments. Now the segments need to be marked for gluing just as in Figure 2, reading counterclockwise with the first segment or side marked for a waist, the second marked *a* and arrowed to the right for gluing, the third marked leg 1, the fourth marked *a* and arrowed to the left for gluing, the fifth marked for a waist, the sixth marked *b* and arrowed to the right for gluing, the seventh marked leg 2, and the eighth marked *b* and arrowed to the left for gluing. Finally, attach the "glued" sides together with masking tape or some other tape that is easy to remove.

If you do not have a Lenart sphere, you may use a close-fitting knit cap, such as the one made in Chapter 2. Use the brim of the cap as an approximation of the great circle, and divide it into eight sections of equal length, aligning them as discussed above. With knit caps, basting with long stitches along the "glued" edges works well.

Carefully observe your result, noting how it both is and is not like a pair of pants that would fit a human. What are the flaws?

Modeling Constant Negative Curvature. A parenthetical remark in Section 2.3 indicates a knitting pattern for making a hyperbolic plane with reduction ratio *n* to $n-1$. The plane is worked from the outside in by casting on a large number of stitches and working the pattern *K($n-2$) K2tog* repeated across each row. In order to space out the decrease stitches somewhat evenly, one must vary the number of stitches one chooses to do before the first K2tog on a row (though this number must be less than $n-2$). As noted earlier, this pattern is inverse to the crochet method designed by Daina Taimina. Using a piece of hyperbolic space made using one of these two methods, Taimina's or ours, we will now mark off a regular octagon on the surface.

Step 1: Begin by choosing a point near the center of the surface to be the center of the octagon.

Step 2: Fold the plane through that point. In folding the plane and flattening along the fold, you will notice that the fold does not form a straight line. The curve it forms is called a *geodesic*. This is truly a line on the hyperbolic plane. Using a piece of thread or yarn, weave in and out along the fold in order to highlight the geodesic.

Step 3: Fold the geodesic made in Step 2 back on itself, through the point you chose in Step 1. This should form a 90° angle very near the point, though the angle will appear to change farther away from the point because of the curvature of the surface. This new fold will make a new geodesic, which you should now mark by weaving.

Step 4: Weave two more geodesics—these will be the angle bisectors of the geodesics you just made.

Step 5: There are now eight radii emerging at equal angles from the center point. Choose a fixed length, and mark off that length from the center on each radius.

Step 6: Connect the points just marked on adjacent radii along geodesics. This forms the regular octagon.

Questions

Why did this construction work? What was the purpose of each step? What were the angles? Why did the angles work out at the center point, even though they didn't work out away from the center? (Hint: Think manifold.) Now that you've done this, you could conceivably make a knitting pattern for a regular octagon. Would you want to do that? Why or why not?

Note to Instructors

One of the authors found this exercise to be very enlightening, but also quite time consuming, so be sure to allow sufficient time—at least an hour for an efficiently working class. Further, use a good contrast color when weaving and small enough stitches so that the geodesics retain their shape after unfolding. One excellent outcome of this problem is that the impetus for the chapter pattern becomes much more clear. Simultaneously, the pattern seems to become much more ingenious.

3.3 Gray Matter and Synapses

The purpose of this section is to investigate and understand the mathematics behind the pattern design in Section 2.5. First we will analyze the general hyperbolic pants pattern construction. Consider Figure 13 and recall that the waist measurement is w in stitches, the cuff measurement is c in stitches, and the waist-to-waist measurement is r in rows and s in stitches.

1. Why should the the number of cast-on stitches be $w/2 + 2s + 2c$, the number of bind-off stitches be $w/2$, and the number of rows be r?

2. Why does this determine the stitch ratio?

3. How many decrease stitches are made in the second row?

4. Why does this mean that the number of stitches in the second row of the pattern is

$$\left\lceil (w/2 + 2s + 2c)\left(\frac{n-1}{n}\right) \right\rceil ?$$

5. Why does this mean that the number of stitches in the third row of the pattern is

$$\left\lceil \left\lceil (w/2 + 2s + 2c)\left(\frac{n-1}{n}\right) \right\rceil \left(\frac{n-1}{n}\right) \right\rceil ?$$

6. Why is the continuous approximation $e^k \approx (n-1)/n$ appropriate?

7. Work out the algebra to figure out why the equation

$$e^k = \left(\frac{w/2}{w/2 + 2s + 2c}\right)^{1/(r-1)}$$

accurately depicts the behavior of the stitching. Note that the formula must go through the cast-on row (row 0, with $w/2 + 2s + 2c$ stitches) and the last row (now row $r - 1$, with $w/2$ stitches), while behaving properly as described in the previous items.

8. Why was it important that we make the continuous approximation rather than using $(n-1)/n$?

Now we will construct a new pattern. Suppose that you have a friend (as we do) whose child has a waist measurement of 19″ and a waist-to-waist measurement of 13″. These measurements are just far enough off of average that any pair of pants from Section 4 will fit poorly. (Of course, the child will have completely different measurements by the time this book is printed, so if you are going to actually knit baby pants instead of just creating a pattern, you may wish to custom-measure a handy baby.)

1. Using a gauge of 5 stitches per inch and 7 rows per inch, compute w, c, r, and s. (*Hint*: You will need to supply some additional information. *Extra hint*: If you are planning to knit pants, buy the yarn you'll use, knit a gauge swatch, and use that gauge instead.)

2. Given the equation $e^k = \left(\dfrac{w/2}{w/2 + 2s + 2c}\right)^{1/(r-1)}$, solve for e^k.

3. From your value for e^k, using trial and error, solve for n. This value of n is an approximate indication of the stitch ratio you actually want. Is this value greater or less than the actual stitch ratio? Given its value, the number of cast-on stitches, and the number of rows, by how much do you predict the approximation for n could be off from the true value?

4. Now use trial and error (and *Mathematica* or some other iterating program on a computer or your calculator) to figure out the actual ratio needed.

5. Use this information to write your personalized hyperbolic pants pattern.

   Good Job!

4   How to Make Hyperbolic Pants

A few notes on the instructions are in order.

⋆ None of these patterns are scalable, and they are highly dependent on the gauges for fit.

⋆ These patterns are forgiving of small mistakes. That is, if you occasionally K40 before a decrease rather than K41, it won't matter. Just be careful near the bind-off row to make sure you have the desired number of stitches.

⋆ Where K2tog or P2tog is marked, feel free to do whatever decrease you find quickest and easiest; it doesn't matter in terms of the finished product. Knitting/purling into the fronts/backs of the loops is fine.

We provide a range of sizes of hyperbolic pants, from infant through five-year-old. Each size has a separate set of instructions. To select the appropriate size, consult Table 1, which contains the measurements of the various finished items. Figure 15 shows several views of the six-month hyperbolic pants in action.

Size	Waist (inches)	Outseam (inches)
newborn	14.5	10
3 month	16	13
6 month	17.5	14.5
9 month	18.5	16
17 month	19.5	17
toddler	20.5	19
5-year-old	22	23

Table 1. A table of sizes.

Figure 15. Three views of the hyperbolic baby pants.

4.1   Ornamental Pants

These pants, shown in Figure 16, will not fit anyone you know—not even a doll (unless, perhaps, it's a Micronaut)! But they are quick to make and illustrate the construction for wearable pants.

Figure 16. Only a Micronaut can wear the ornamental pants.

Materials

One ball of yarn and appropriately sized needles. For firmer fabric than usual, use needles one to two sizes smaller than recommended.

Instructions

Cast on 63 stitches using a long-tail cast-on.

Row 1: *Purl 3, Purl 2 together.* Repeat from * across. You might not end at the end of *, but just end at the end of the row.

Row 2: Knit 1, Knit 2 together. *Knit 3, Knit 2 together.* Repeat from * across, until all stitches are completed.

Row 3: Purl 2, Purl 2 together. *Purl 3, Purl 2 together.* Repeat from * across, until all stitches are completed.

Row 4: Knit 2, Knit 2 together. *Knit 3, Knit 2 together.* Repeat from * across, until all stitches are completed.

Row 5: Purl 2, Purl 2 together. *Purl 3, Purl 2 together.* Repeat from * across, until all stitches are completed.

Row 6: Knit 1, Knit 2 together. *Knit 3, Knit 2 together.* Repeat from * across, until all stitches are completed.

Row 7: Purl 1, Purl 2 together. *Purl 3, Purl 2 together.* Repeat from * across, until all stitches are completed.

Row 8: Knit 1, Knit 2 together. *Knit 3, Knit 2 together.* Repeat from * across, until all stitches are completed.

Row 9: Purl 2, Purl 2 together. *Purl 3, Purl 2 together.* Repeat from * across, until all stitches are completed.

Row 10: Knit 3, Knit 2 together. *Knit 3, Knit 2 together.* Repeat from * across, until all stitches are completed.

Row 11: Purl 2, Purl 2 together. *Purl 3, Purl 2 together.* Repeat from * across, until all stitches are completed.

Row 12: Knit 1, Knit 2 together. *Knit 3, Knit 2 together.* Repeat from * across, until all stitches are completed.

Row 13: Bind off.

Sew the Outseams

Count 29 stitches in from one edge of the cast-on and place a marker. From this point, count over the same number as your bound-off stitches (5 or 6) and place a second marker. The bound-off stitches will form one half of the waist, and the stitches between the markers will form the other half of the waist. Fold the pants over, wrong sides together, so that the halves of the waist match. Using an invisible join such as the vertical-to-horizontal graft, seam the selvedge to the corresponding 15 stitches of the cast-on edge on each side.

4.2   Newborn Pants

Be sure to check the notes at the start of Section 4. Finished newborn pants are pictured in Figure 17.

Materials

Two 50g balls (250 m) of light worsted-weight, machine-washable yarn that gives a gauge of 5 stitches per inch and 7 rows per inch. In this pattern, the row gauge is as important as the stitch gauge. Possible yarns include Valley Superwash with size 5 needles and Cascade Bollicine Holiday with size 6 needles.

Figure 17. Completed newborn pants.

Instructions

Using a long-tail cast-on, cast on 206 stitches. Place split stitch markers on the cast-on edge (not the needle) between the 35th and 36th stitches, 85th (= 35 + 50) and 86th stitches, 121st (= 35 + 50 + 36) and 122nd stitches, and 171st (= 35 + 50 + 36 + 50) and 172nd stitches. These will be used as guides for seaming the pants and must be placed before any further knitting occurs to avoid difficulty in counting once decreases have been made.

Row 1: *P30, P2tog*, finishing the row when the stitches run out.

Row 2: *K30, K2tog*, finishing the row when the stitches run out.

Row 3: P25, *P2tog, P30*, finishing the row when the stitches run out.

Row 4: K25, *K2tog, K30*, finishing the row when the stitches run out.

Repeat this pattern 17 times, for a total of 68 rows (or until there are 36 stitches on the needles).

Row 69: Purl across.

Bind off loosely purlwise.

At this point your pants will not look like pants, but instead like a piece of a pseudosphere as shown in Figure 18.

Sew the Outseams

With the knitting right-side out, match the middle two stitch markers (on the cast-on edge) with the two ends of the bind-off edge. Use an invisible vertical-to-horizontal graft (such as in [9, p. 100]) to seam the cast-on edge to each selvage. These seams begin at the waist, where the bind-off and cast-on stitches are matched, and end at the cuffs, where the remaining stitch markers meet the corner of the selvage and cast-on edge. Every three stitches, graft to two rows instead of one; for a more precise matching, use the algorithm given for picking up stitches in Chapter 2.

Thread 15″ of thin elastic or two 30″ lengths of yarn through every other stitch just below the cast-off edge. Sew the ends of the elastic together, or tie the lengths of yarn into a bow.

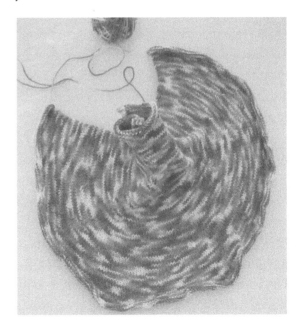

Figure 18. Baby pants in the snow before they've been sewn up.

4.3 Three Month Baby Pants

Be sure to check the notes at the start of Section 4. Finished three-month-old baby pants are pictured in Figure 19.

Figure 19. Completed three-month-old baby pants.

Materials

Two 50g balls (250 m) of light worsted-weight, machine-washable yarn that gives a gauge of 5 stitches per inch and 7 rows per inch. In this pattern, the row gauge is as important as the stitch gauge. Possible yarns include Valley Superwash with size 5 needles and Cascade Bollicine Holiday with size 6 needles.

Instructions

Using a long-tail cast-on, cast on 250 stitches. Place split stitch markers on the cast-on edge (not the needle) between the 40th and 41st stitches, 105th (= 40 + 65) and 106th stitches, 145th (= 40 + 65 + 40) and 146th stitches, and 210th (= 40 + 65 + 40 + 65) and 211th stitches. These will be used as guides for seaming the pants and must be placed before any further knitting occurs to avoid difficulty in counting once decreases have been made.

Row 1: *P38, P2tog*, finishing the row when the stitches run out.

Row 2: *K38, K2tog*, finishing the row when the stitches run out.

Row 3: P32, *P2tog, P38*, finishing the row when the stitches run out.

Row 4: K32, *K2tog, K38*, finishing the row when the stitches run out.

Repeat this pattern 22 times, for a total of 88 rows (or until there are 42 stitches on the needles).

Row 89: *P38, P2tog*, finishing the row when the stitches run out.

Row 90: Knit across.

Bind off loosely knitwise.

At this point your pants will not look like pants, but instead like a piece of a pseudosphere as shown in Figure 18.

Sew the Outseams

With the knitting right-side out, match the middle two stitch markers (on the cast-on edge) with the two ends of the bind-off edge. Use an invisible vertical-to-horizontal graft (such as in [9, p. 100]) to seam the cast-on edge to each selvage. These seams begin at the waist, where the bind-off and cast-on stitches are matched, and end at the cuffs, where the remaining stitch markers meet the corner of the selvage and cast-on edge. Every three stitches, graft to two rows instead of one; for a more precise matching, use the algorithm given for picking up stitches in Chapter 2.

Thread 16″ of thin elastic or two 32″ lengths of yarn through every other stitch just below the cast-off edge. Sew the ends of the elastic together, or tie the lengths of yarn into a bow.

4.4 Six Month Baby Pants

Be sure to check the notes at the start of Section 4. Finished six-month-old baby pants are pictured in Figure 20.

Materials

Three 50g balls (300 m) of worsted-weight, machine-washable yarn that gives a gauge of 5 stitches per inch and 6.5 rows per inch. In this pattern, the row gauge is as important as the stitch gauge. Possible yarns include Cascade 220 Superwash and Cascade Bollicine Maxi with size 6 needles.

Instructions

Using a long-tail cast-on, cast on 280 stitches. Place split stitch markers on the cast-on edge (not the needle) between the 45th and 46th stitches, 118th (= 45 + 73) and 119th stitches, 162nd (= 45 + 73 + 44) and 163rd stitches, and 234th (= 45 + 73 + 44 + 72) and 235th stitches. These will be used as guides for seaming the pants, and must be placed before any further knitting occurs to avoid difficulty in counting once decreases have been made.

Row 1: *P39, P2tog*, finishing the row when the stitches run out.

Row 2: *K39, K2tog*, finishing the row when the stitches run out.

Row 3: P34, *P2tog, P39*, finishing the row when the stitches run out.

Row 4: K34, *K2tog, K39*, finishing the row when the stitches run out.

Repeat this pattern 23 times, for a total of 92 rows (or until there are 44 stitches on the needles).

Row 93: Purl across.

Bind off loosely purlwise.

At this point your pants will not look like pants, but instead like a piece of a pseudosphere as shown in Figure 18.

Sew the Outseams

With the knitting right-side out, match the middle two stitch markers (on the cast-on edge) with the two ends of the bind-off edge. Use an invisible vertical-to-horizontal graft (such as in [9, p. 100]) to seam the cast-on edge to each selvage. These seams begin at the waist, where the bind-off and cast-on stitches are matched, and end at the cuffs, where the remaining stitch markers meet the corner of the selvage and cast-on edge. Every three or four stitches, graft to two rows

instead of one; for a more precise matching, use the algorithm given for picking up stitches in Chapter 2.

Thread 18″ of thin elastic or two 35″ lengths of yarn through every other stitch just below the cast-off edge. Sew the ends of the elastic together, or tie the lengths of yarn into a bow.

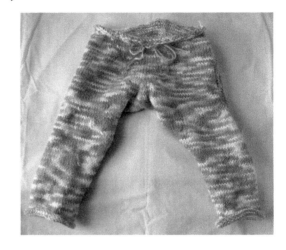

Figure 20. Completed six-month-old baby pants.

4.5 Nine Month Baby Pants

Be sure to check the notes at the start of Section 4.

Materials

Four 50g balls (∼330 m) of worsted-weight, machine-washable yarn that gives a gauge of 5 stitches per inch and 6.5 rows per inch. In this pattern, the row gauge is as important as the stitch gauge. Possible yarns include Cascade 220 Superwash and Cascade Bollicine Maxi with size 6 needles.

Instructions

Using a long-tail cast-on, cast on 306 stitches. Place split stitch markers on the cast-on edge (not the needle) between the 50th and 51st stitches, 130th (= 50 + 80) and 131st stitches, 176th (= 50 + 80 + 46) and 177th stitches, and 256th (= 50 + 80 + 46 + 80) and 257th stitches. These will be used as guides for seaming the pants, and must be placed before any further knitting occurs to avoid difficulty in counting once decreases have been made.

Row 1: *P42, P2tog*, finishing the row when the stitches run out.

Row 2: *K42, K2tog*, finishing the row when the stitches run out.

Row 3: P37, *P2tog, P42*, finishing the row when the stitches run out.

Row 4: K37, *K2tog, K42*, finishing the row when the stitches run out.

Repeat this pattern 25 times, for a total of 100 rows (or until there are 48 stitches on the needles).

Row 101: *P42, P2tog*, finishing the row when the stitches run out.

Row 102: *K42, K2tog*, finishing the row when the stitches run out.

Row 103: Purl across.

Bind off loosely purlwise.

At this point your pants will not look like pants, but instead like a piece of a pseudosphere as shown in Figure 18.

Sew the Outseams

With the knitting right-side out, match the middle two stitch markers (on the cast-on edge) with the two ends of the bind-off edge. Use an invisible vertical-to-horizontal graft (such as in [9, p. 100]) to seam the cast-on edge to each selvage. These seams begin at the waist, where the bind-off and cast-on stitches are matched, and end at the cuffs, where the remaining stitch markers meet the corner of the selvage and cast-on edge. Every three or four stitches, graft to two rows instead of one; for a more precise matching, use the algorithm given for picking up stitches in Chapter 2.

Thread 19″ of thin elastic or two 39″ lengths of yarn through every other stitch just below the cast-off edge. Sew the ends of the elastic together, or tie the lengths of yarn into a bow.

4.6 Seventeen Month Baby Pants

Be sure to check the notes at the start of Section 4.

Materials

Four 50g balls (∼300 yds) of heavy worsted-weight, machine-washable yarn that gives a gauge of 4 stitches per inch and 5.5 rows per inch. In this pattern, the row gauge is as important as the stitch gauge. One possible yarn is Laines du Nord Dolly Maxi with size 9 needles.

Instructions

Using a long-tail cast-on, cast on 267 stitches. Place split stitch markers on the cast-on edge (not the needle) between the 46th and 47th stitches, 114th (= 46 + 68) and 115th stitches, 153rd (= 46 + 68 + 39) and 154th stitches, and 221st (= 46 + 68 + 39 + 68) and 222nd stitches. These will be used as guides for seaming the pants, and must be placed before any further knitting occurs to avoid difficulty in counting once decreases have been made.

Row 1: *P38, P2tog*, finishing the row when the stitches run out.

Row 2: *K38, K2tog*, finishing the row when the stitches run out.

Row 3: P34, *P2tog, P38*, finishing the row when the stitches run out.

Row 4: K34, *K2tog, K38*, finishing the row when the stitches run out.

Repeat this pattern 23 times, for a total of 92 rows (or until there are 39 stitches on the needles).

Row 93: Purl across.

Bind off loosely purlwise.

At this point your pants will not look like pants, but instead like a piece of a pseudosphere as shown in Figure 18.

Sew the Outseams

With the knitting right-side out, match the middle two stitch markers (on the cast-on edge) with the two ends of the bind-off edge. Use an invisible vertical-to-horizontal graft (such as in [9, p. 100]) to seam the cast-on edge to each selvage. These seams begin at the waist, where the bind-off and cast-on stitches are matched, and end at the cuffs, where the remaining stitch markers meet the corner of the selvage and cast-on edge. Every three or four stitches, graft to two rows instead of one; for a more precise matching, use the algorithm given for picking up stitches in Chapter 2.

Thread 20″ of thin elastic or two 34″ lengths of yarn through every other stitch just below the cast-off edge. Sew the ends of the elastic together, or tie the lengths of yarn into a bow.

4.7 Toddler Pants

Be sure to check the notes at the start of Section 4.

Materials

Four 50g balls (∼340 yds) of heavy worsted-weight, machine-washable yarn that gives a gauge of 4 stitches per inch and 5.5 rows per inch. In this pattern, the row gauge is as important as the stitch gauge. One possible yarn is Laines du Nord Dolly Maxi with size 9 needles.

Instructions

Using a long-tail cast-on, cast on 293 stitches. Place split stitch markers on the cast-on edge (not the needle) between the 50th and 51st stitches, 126th (= 50 + 76) and 127th stitches, 167th (= 50 + 76 + 41) and 168th stitches, and 243rd (= 50 + 76 + 41 + 76) and 244th stitches. These will be used as guides for seaming the pants, and must be placed before any further knitting occurs to avoid difficulty in counting once decreases have been made.

Row 1: *P40, P2tog*, finishing the row when the stitches run out.

Row 2: *K40, K2tog*, finishing the row when the stitches run out.

Row 3: P35, *P2tog, P40*, finishing the row when the stitches run out.

Row 4: K35, *K2tog, K40*, finishing the row when the stitches run out.

Repeat this pattern 25 times, for a total of 100 rows (or until there are 44 stitches on the needles).

Row 101: *P40, P2tog*, finishing the row when the stitches run out.

Row 102: *K40, K2tog*, finishing the row when the stitches run out.

Row 103: P35, *P2tog, P40*, finishing the row when the stitches run out.

Row 104: Knit across.

Bind off loosely knitwise.

At this point your pants will not look like pants, but instead like a piece of a pseudosphere as shown in Figure 18.

Sew the Outseams

With the knitting right-side out, match the middle two stitch markers (on the cast-on edge) with the two ends of the bind-off edge. Use an invisible vertical-to-horizontal graft (such as in [9, p. 100]) to seam the cast-on edge to each selvage. These seams begin at the waist, where the bind-off and cast-on stitches are matched, and end at the cuffs, where the remaining stitch markers meet the corner of the selvage and cast-on edge. Every three or four stitches, graft to two rows instead of one; for a more precise matching, use the algorithm given for picking up stitches in Chapter 2.

Thread 21″ of thin elastic or two 41″ lengths of yarn through every other stitch just below the cast off edge. Sew the ends of the elastic together, or tie the lengths of yarn into a bow.

4.8 Children's Pants

Be sure to check the notes at the start of Section 4.

Materials

Six 50g balls (∼400 yds) of bulky, machine-washable yarn that gives a gauge of 3.5 stitches per inch and 5 rows per inch. In this pattern, the row gauge is as important as the stitch gauge. One possible yarn is Jaeger Extra Fine Merino Chunky with size 10 needles.

Instructions

Using a long-tail cast-on, cast on 298 stitches. Place split stitch markers on the cast-on edge (not the needle) between the 49th and 50th stitches, 130th (= 49 + 81) and 131st stitches, 169th (= 49 + 81 + 39) and 170th stitches, and 249th (= 49 + 81 + 39 + 80) and 250th stitches. These will be used as guides for seaming the pants, and must be placed before any further knitting occurs to avoid difficulty in counting once decreases have been made.

Row 1: *P38, P2tog*, finishing the row when the stitches run out.

Row 2: *K38, K2tog*, finishing the row when the stitches run out.

Row 3: P34, *P2tog, P38*, finishing the row when the stitches run out.

Row 4: K34, *K2tog, K38*, finishing the row when the stitches run out.

Repeat this pattern 28 times, for a total of 112 rows (or until there are 40 stitches on the needles).

Row 101: *P38, P2tog*, finishing the row when the stitches run out.

Row 103: Knit across.

Bind off loosely knitwise.

At this point your pants will not look like pants, but instead like a piece of a pseudosphere as shown in Figure 18.

Sew the Outseams

With the knitting right-side out, match the middle two stitch markers (on the cast-on edge) with the two ends of the bind-off edge. Use an invisible vertical-to-horizontal graft (such as in [9, p. 100]) to seam the cast-on edge to each selvage. These seams begin at the waist, where the bind-off and cast-on stitches are matched, and end at the cuffs, where the remaining stitch markers meet the corner of the selvage and cast-on edge. Every three or four stitches, graft to two rows

instead of one; for a more precise matching, use the algorithm given for picking up stitches in Chapter 2.

Thread 22″ of thin elastic or two 44″ lengths of yarn through every other stitch just below the cast-off edge. Sew the ends of the elastic together, or tie the lengths of yarn into a bow.

Bibliography

[1] belcastro, sarah-marie, and Yackel, Carolyn. *Mathematical Knitting Patterns*. In preparation.

[2] do Carmo, Manfredo P. *Differential Geometry of Curves and Surfaces.* Prentice-Hall, Inc., Englewood Cliffs, NJ, 1976.

[3] Doyle, Peter. "Curvature of Surfaces." *The Geometry Center.* http://www.geom.uiuc.edu/docs/doyle/mpls/handouts/node21.html, April 12, 1994.

[4] Henderson, David, and Taimina, Daina. "Crocheting the Hyperbolic Plane." *Mathematical Intelligencer*, vol. 23, no. 2, Spring 2001, pp. 17–28.

[5] Rucker, Rudolf v. B. *Geometry, Relativity and the Fourth Dimension.* Dover, New York, 1977.

[6] Singer, I. M., and Thorpe, John A. *Lecture Notes on Elementary Topology and Geometry.* Scott, Foresman and Company, Glenview, IL, 1967.

[7] Taimina, Daina. *Exploring Non-Euclidean Geometry Through Crochet.* To appear.

[8] Taimina, Daina. http://www.math.cornell.edu/~dtaimina/, 2006.

[9] Vogue Knitting Magazine Editors. *Vogue Knitting: The Ultimate Knitting Book.* Sixth&Spring, New York, 2002.

ABOUT THE CONTRIBUTORS

The Editors have done quite a variety of things connecting mathematics and fiber arts. sarah-marie focuses on knitting, but has designed a number of different objects and continues to challenge herself to knit new objects and answer new questions in doing so. Carolyn finds mathematics in a number of different media—knitting, crocheting, tatting, temari balls. We have both made a distinct effort over a number of years to foster a mathematical community of fiber artists. The idea of the community has been to encourage mathematicians/fiber artists to notice and articulate the mathematics in their own work. We have done this by holding the Knitting Network at the AMS-MAA Joint Mathematics Meetings since 2000, at MathFest, and at the 2006 AP Calc reading.

Amy F. Szczepański learned to knit and crochet from her grandmother when she was in elementary school and picked up embroidery after a stint in the Girl Scouts. While an undergraduate at Dartmouth College, her design course included a project based on traditional patchwork quilting, and, armed with her grandmother's sewing machine, she was inspired to teach herself quilting. Amy earned her Ph.D. in mathematics at the University of California San Diego and now teaches at the University of Tennessee. She lives in Knoxville, TN where, with her husband Jim Conant, she is restoring a 1920 bungalow.

Carolyn Yackel teaches mathematics at Mercer University, where she has developed a general-education course in mathematics through fiber arts and regularly teaches for the Interdisciplinary Studies program. She has more interests than any ten normal people put together. For example, Carolyn makes both functional and artistic pottery, loves to garden, and is an avid self-taught cook specializing in desserts and international foods. Carolyn did her undergraduate work at the University of Chicago and her Ph.D. at the University of Michigan. Her research areas include commutative algebra and mathematics education.

D. Jacob (Jake) Wildstrom is finishing his graduate work in mathematics at the University of California, San Diego. His research areas include dynamic optimization, integer programming, facility location, and games on graphs. He was introduced to crochet by a textile-arts seminar co-sponsored by Epsilon Theta Fraternity and the Experimental Study Group of MIT in 2001. He enjoys Hungarian culture, board games, interactive fiction, and graphic novels. His crafting ambitions include rendering the board game *The Settlers of Catan* in crochet and integrating graph-theoretical concepts into his work.

Joshua Holden, now at Rose-Hulman Institute of Technology, received his Ph.D. from Brown University. His research interests are in computational and algebraic number theory, cryptography, and the application of graph theory to fiber arts. His fiberological interests include blackwork (of course), counted-canvas embroidery, hardanger, and crochet. He is interested in the use of technology in teaching and the use of historically informed pedagogy. Still largely in the non-mathematical category are his interests in science fiction and music, both classical and contemporary.

Lana Holden holds the degree of Artium Magistri from Dartmouth College but considers herself a "recovering mathematician" in that she insistently declares knitting to be her passion and math to be a sideline. She teaches fiber arts classes, including the popular "Math for Knitters," at River Wools, a yarn store in Terre Haute, IN. She enjoys knitting and designing patterns with unusual geometric construction techniques and credits Norah Gaughan as her primary inspiration. In her spare time, Lana plays keyboard and sings with the classic alternative band Whisper Down.

Mary Shepherd teaches at Northwest Missouri State University. In addition to her Ph.D. in Mathematics from Washington University in St. Louis, Mary has a Bachelor of Music degree from Missouri State University and a Masters of Accountancy from the University of Oklahoma. She learned cross-stitching so she could teach it at Better Homes and Gardens Craft Creations parties (like Tupperware parties). Mary also runs 5Ks during the summer months and performs on clarinet throughout the year with the Cameron Municipal Band. She is married and has two boys, Christopher and Brent.

sarah-marie belcastro co-directs the Hampshire College Summer Studies in Mathematics (www.hcssim.org) and is currently a Visiting Assistant Professor of Mathematics at Smith College. Among her many not-pure-mathematics interests are the feminist philosophy of science, dance (principally ballet and modern), her (17+)-year-old cat, and changing the world. sarah-marie's primary mathematical research area is topological graph theory. She did her undergraduate work at Haverford College and her Ph.D. at the University of Michigan. For more, see her domain at www.toroidalsnark.net.

Susan Goldstine enjoys making tactile and visual mathematical models, employing such diverse media as yarn, fabric, thread, glass beads, paper, steel wire, copper tubes, pinecones, and pottery, though not all at once. She is an avid cook, and hopes one day to reproduce the interlocking Escher swan cookies she made as an undergraduate. Susan received her BA in Mathematics and French from Amherst College and her Ph.D in Mathematics from Harvard University. Her current institutional home is St. Mary's College of Maryland, where she strives to maintain her reputation for having the office with the most toys.

CREDITS

"THINKING INSIDE THE BOX"
2001, 55"X55"
JERI RIGGS

Anything not credited has been created and/or photographed by the chapter author.

Introduction. The opening figure: temari balls representing Platonic solids, made by Carolyn Yackel.

Chapter 1. The opening figure: quilts by Jeri Riggs, photograph by sarah-marie belcastro. Figures 15, 16, 17, and 18 on pages 25, 26, 27, and 27: quilts and photographs by Jeri Riggs. Diagrams created by Amy F. Szczepański and sarah-marie belcastro using Adobe Illustrator and ACD Systems Canvas.

Chapter 2. The opening figure and Figure 7 on page 37: hat in Noro Silk Garden, modeled by Ashley Hatfield, photograph by sarah-marie belcastro. Figure 2 on page 31: sweater by Jeri Carroll, photograph by Martha Crossen. Figures 3, 4, and 9 on pages 32, 32, and 38: modeled by Andrea Layton. Figure 5 on page 33: hat in Melody color 02 by Yarn TreeHouse (eBay seller) and photograph by sarah-marie belcastro, modeled by Joan Belcastro. Figure 8 on page 37: hats and photographs by sarah-marie belcastro; Noro Blossom hat modeled by Frank Belcastro and Berroco Air (Cyberscape; less than one ball!) hat modeled by Rachel Brown. Diagrams created by sarah-marie belcastro and Lana Holden using ACD Systems Canvas.

Chapter 3. The opening figure: shawl by Ida Thornton, modeled by Hope Stansfield, photograph by sarah-marie belcastro in the Smith College Botanic Garden. Figure 5 on page 45: shawl in Misti Alpaca Lace Merlot, Maize, and Cranberry by Em Schoch, modeled by Zia Marek-Loftus, photographs by sarah-marie belcastro. Figures 8, 10, and 12 on pages 47, 50, and 51: shawl by D. Jacob Wildstrom, photographs by sarah-marie belcastro. Modeled by Hope Stansfield in Figure 10. Figure 11 on page 51: miniature Sierpinski shawl by Carolyn Yackel, modeled by Patricia Williams, photograph by Creighton Rosental. Diagrams created using Wolfram Mathematica and MetaPost.

Chapter 4. The opening figure and Figure 12 on page 66: tori in Reynolds Saucy Sea Foam and Reynolds Lite-Lopi color 0304 by Zia Marek-Loftus. Figures 8 and 13 on pages 63 and 68: orange torus Classic Elite Flash in Clementine, blue/grey/purple torus in Berrocco Tuscany color 1512, white/pastel torus in Filatura di Crosa Lovely, cream/blue torus in Southwool Handpainted (from Uruguay via eBay), cabled torus in Filatura di Crosa Brilla, and blue/green torus in Reynolds Garden Tweed color 09. The feline model is Arzachel (http://www.toroidalsnark.net/RZ.html). Diagrams created in ACD Systems Canvas.

Chapter 5. The opening figure: pillow by Zia Marek-Loftus, photograph by sarah-marie belcastro. Figure 22 on page 84: work-in-progress and photograph by Susan Sierra. Figures 23, 24, 25, and 26 on pages 85, 86, 87, and 88: produced using Cross Stitch Professional for Windows (www.dpsoftware.co.uk). Figures 5–12 on pages 73–74: redrawn in Adobe Illustrator by Charlotte Henderson. Cross-stitch directions were modified from those with Design Works Crafts pattern for Shakespeare's Garden 9889 and Dimensions kit #16657 A Gift of Love Birth Record.

Chapter 6. Figures 3 and 5 on pages 94 and 98: Lavender and lime socks in Classic Elite Four Season Cotton colors Spring Green and King's Purple by Ashley Hatfield, modeled by Zia Marek-Loftus, photographs by sarah-marie belcastro. The opening figure and Figure 7 on page 100: Navy and white socks by sarah-marie belcastro and Joan M. Gallant Belcastro, modeled by Joan M. Gallant Belcastro, photographs by sarah-marie belcastro. Diagrams created by sarah-marie belcastro using ACD Systems Canvas.

Chapter 7. Figure 1 on page 106: Klein bottle is from Acme Klein Bottle Company, http://www.kleinbottle.com/. Figure 8 on page 111: quilt by Jeri Riggs, photograph by sarah-marie belcastro. Jeri Riggs studied mathematics and psychology at MIT and Wellesley College and earned her M.D. from Boston University. Since retiring from practicing psychiatry, she has enjoyed making quilts that incorporate her love of color, geometry and line. You can see more of her quilts at http://www.jeririggs.com and follow her work in progress at http://jeririggs.blogspot.com. Figure 14 on page 115: modeled by Portia Parker and Madison Stuart, photographs by sarah-marie belcastro. Diagrams created in Adobe Illustrator and Photoshop.

Chapter 8. Figure 1 on page 119: sweaters owned by Tom Hull, Jim Henle, and Sean Kinlin, respectively, photographs by sarah-marie belcastro. Figure 2 on page 120: modeled by Sean Kinlin, photograph by sarah-marie belcastro. Figure 5 on page 122: sweaters owned by Tom Hull and Sean Kinlin, respectively, photographs by sarah-marie belcastro. Figure 10 on page 126: sweaters owned by Tom Hull, sarah-marie belcastro, and Rachel Brown, respectively, photographs by sarah-marie belcastro. Figure 8 on page 127: pillow in Valley Goshen Sage by Zia Marek-Loftus, photographs by sarah-marie belcastro. Figure 13 on page 128: pillow in Blue Sky Organic Cotton Sage and photographs by Carolyn Yackel. Figure 14 on page 129: hanging in Lion Brand Cotton Ease Stone and photograph by Charlotte Davis. Diagrams created by sarah-marie belcastro using ACD Systems Canvas.

Chapter 9. All photographs except Figures 13 and 14 are by Andrea Layton. The opening figure: stitched on 32 count Belfast linen using one strand of dark green silk by Ann Black. Figure 6 on page 139: stitched by Ann Black. Figure 10 on page 142: stitched by Lana Holden. Figures 13 and 14 on pages 149 and 150: motif stitched on a Taylor & Francis complimentary Joint Mathematics Meetings tote bag by Kim Plofker, photographs by Toke Lindegaard Knudsen.

Chapter 10. The opening figure and Figures 14 and 15 on pages 165 and 169: pants by sarah-marie belcastro, modeled by Michela Rowland, photographs by Patrick Rowland and Thomas Hull. Figures 16 and 17 on pages 170 and 171: pants knitted by Zia Marek-Loftus, Micronaut contributed by Thomas Hull, photographs by sarah-marie belcastro. Figure 19 on page 172: pants knitted by Ashley Hatfield, photograph by sarah-marie belcastro. Diagrams created by sarah-marie belcastro using ACD Systems Canvas.

INDEX

Teaching Venues

We list here pointers to places in the book appropriate for particular subjects or venues.

Making Mathematics with Needlework

Designed by Erica Schultz
Typeset by A K Peters, Ltd.
Printed and bound by Replika Press, Pvt, India

Composed in Adobe Myriad using LaTeX
Printed on 90 gsm Gloss Art Paper